# Lysosomes and Cell Function

# Integrated Themes in Biology

**Consulting Editor:** I. D. J. Phillips, University of Exeter

**Pitt:** Lysosomes and Cell Function
**Bradbeer and Bradbeer:** The Regulation of Metabolism

# Lysosomes and Cell Function

**D. Pitt**

Department of Biological Sciences
University of Exeter
England

Longman   LONDON and NEW YORK

**Longman Group Limited**
Burnt Mill
Harlow
Essex CM20 2JE

Published in the United States of America by
Longman Inc., New York

*Associated companies, branches and representatives
throughout the world*

*First published,* 1975

ISBN 0 582 44344 X

*Printed in Great Britain by
Whitstable Litho Ltd*

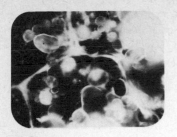

# Preface

During the period that my research interests have centred on lysosomes I have often been asked by both students and teachers of cell biology to recommend a general text on lysosomes for undergraduate courses. Although the research literature pertaining to lysosomes is extensive and has recently been summarized in three large volumes, there is presently no introductory text available. Consequently, the student belief is that lysosomes are recherché organelles with an abstruse, mainly medical, literature and merit scarcely a paragraph in most general texts on cell biology. Intensified research effort in studying lysosomes has now established these organelles as part of a dynamic system which is of major importance to cells and organisms. I felt it was time to assemble these important findings in a general student text which attempts to convey the lysosome concept in the widest context of biology.

Although lysosomes were first discovered by medical science it is now believed that they are of almost ubiquitous occurrence in organisms, consequently I have sought to provide a wide coverage of their distribution and functions in animals, plants and protists. Since our knowledge of some aspects of lysosome function is incomplete, speculative and tentative, I have, on occasions, expressed a personal opinion, but at the same time I have given references in the text, or as suggested reading, which provide access to the sometimes conflicting ideas in the current research literature.

The book is divided into chapters in the traditional way, but I have found it necessary, owing to the integrated nature of the lysosome system, to recapitulate and reiterate facts and information in the various chapters. I feel, therefore, that the chapters form a sextet of interrelated essays.

In order to keep the book to a reasonable size and price, and to fulfil the original objective of providing a general text for students I have had to assume degrees of knowledge of various diverse aspects of biology. Here also I have sought to provide access to appropriate literature which should supply background information.

Since the rapid progress in research on lysosomes is attributable to the multidisciplinary experimental approach adopted by workers in the field I have included an appendix of basic methodology which should be within the scope of most competent students of cell physiology who have access to certain items of routine equipment. Each of these selected methods has been successfully used at some time in my own laboratory and I feel that if the impetus in the study of lysosomes, or indeed all aspects of sub-cellular organization, is to be maintained it is essential that students gain familiarity with and competence in the use of techniques. I hope, therefore, that the appendix will encourage students to experiment with lysosomes.

Finally, I wish to express my thanks to all the research workers whose efforts form the basis of this book, and particularly to those who generously permitted the use of data and illustrative material and provided so many original prints.

# Contents

Preface     v

1   The Lysosome System     1

2   Occurrence of Lysosomes     24

3   Physiological Functions of Lysosomes: Role of Lysosomes in Intracellular Digestion     52

4   Physiological Functions of Lysosomes: Role of Lysosomes in Secretion and Extracellular Digestion     79

5   Role of Lysosomes and Acid Hydrolases in Developmental Processes     95

6   Lysosomes in Disease and Injury     126

Appendix     146

Index     157

# Chapter 1

# The Lysosome System

## Introduction

During the last century cytologists came to recognize sub-cellular organelles including nuclei, mitochondria, chloroplasts and the Golgi apparatus. It was also noted that cells contained numerous refractile granules known as 'lipochondria' as distinct from the superficially similar mitochondria. The study of sub-cellular organization gained new impetus from the important work of Gomori who expounded the principles of enzyme cytochemistry and in 1939 located alkaline phosphatase within tissue sections using the light microscope. Meanwhile biochemists had been making important advances in understanding metabolic pathways, mainly through the use of soluble tissue extracts. The current state of our knowledge relating to these matters testifies to the value and success of this approach. However, the sedimentable fraction of such tissue homogenates was often discarded, even though Warburg had demonstrated, as early as 1913, that respiration was associated with sedimentable particles. In the 1940s classic work by Claude revealed that cells could be broken under controlled conditions and the contents separated into various fractions by differential centrifugation. This technique opened the way for other research groups to show that morphologically recognizable mitochondria were the sub-cellular sites of oxidative phosphorylation. At this time Chantrenne was able to demonstrate that when nuclei and mitochondria were removed from cell homogenates the supernatant fluid fractions contained many smaller particles including the ribosomes and fragments of the endoplasmic reticulum, known collectively as 'microsomes', which could be sedimented at very high centrifugal forces and characterized by electron microscopy.

Using refinements of these various techniques it soon became apparent that the mitochondrial pellet could be sub-fractionated into further particle types which contained arrays of both oxidative and hydrolytic enzymes. In the early 1950s Christian de Duve and his group in Louvain were able to devise methods which resolved the crude mitochondrial pellet from rat liver homogenates into two fractions — purified mitochondria and a further group of particles containing acid hydrolases and initially called the light mitochondrial fraction or L fraction, but which were subsequently named lysosomes (i.e. lytic bodies). During this period pioneering studies were carried out by Novikoff in America when he was able to complement the biochemical work on lysosomes with observations at the light microscope and ultrastructural levels. The lysosomes were, therefore, the first organelles to be so extensively studied by co-ordinated biochemical, cytochemical and cytological techniques. Such a multidisciplinary approach to cell biology is now commonplace and the techniques employed so successfully in these initial studies have merged to form the field of endeavour known today as enzyme cytology whereby biologists seek to determine the specific function(s) of sub-cellular components in the life of the cell. This approach has paid handsome dividends in the past decade and has allowed penetrating insight into the nature of sub-cellular organization — a far cry from the pioneering studies of the Büchner's on yeast juice in 1897.

## 1.1  The Lysosome Concept

Lysosomes were discovered almost twenty years ago in rat liver by de Duve and co-workers in Belgium following extensive work on the cellular location of certain hydrolytic enzymes (de Duve *et al.*, 1955).

Hydrolytic enzymes, or *hydrolases*, are capable of hydrolytic splitting of a wide range of natural substrates including esters, proteins and peptides, carbohydrates, lipids and nucleotides, according to the overall reaction:

$$R-R^1 + H . OH \rightleftharpoons R . H + R^1-OH$$

The earlier work of de Duve's group demonstrated that five such hydrolases, all characterized by *acid pH optima* were sedimentable from carefully-prepared rat-liver homogenates by differential centrifugation techniques, and were associated with a specific *cell particle* approximately 0.2–0.4 $\mu$m in diameter. Additionally, it was found that the enzymes associated with the sedimented particles demonstrated *latency* since they failed to hydrolyse appropriate substrates until they were released from particulate sites by osmotic shock, detergents, lecithinases and freezing and thawing: treatments known to disrupt cell membranes. After such treatments much of the previously particle-bound latent enzyme was

2

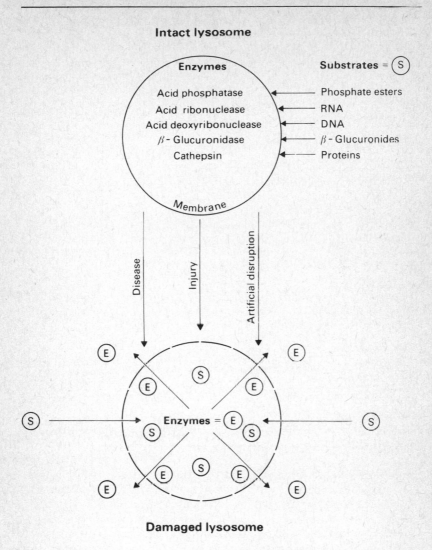

**Fig. 1.1** Early model of the lysosome based on the original ideas of de Duve and co-workers. The membrane in the normal state is impermeable to outward passage of the enclosed enzymes and to inward flow of substrates, S, which must gain access to the interior of the organelle before hydrolysis can occur. Membrane permeability may be altered by various artificial means when the enzymes, E, become 'activated'. During cell injury and disease enzymes may be liberated and cause varying degrees of metabolic disturbance. Only the five enzymes originally associated with the concept are included.

**Table 1.1**   Some acid hydrolases which have been located in lysosomes

| Enzyme | Natural substrate | Source of lysosomes |
|---|---|---|
| *Phosphatases* | | |
| Acid phosphatase | Most phosphomonoesters | Many tissues of |
| Acid phosphodiesterase | Oligonucleotides and phosphodiesters | animals and plants; protists |
| *Nucleases* | | |
| Acid ribonuclease | RNA | Many tissues of |
| Acid deoxyribonuclease | DNA | animals and plants; protists |
| *Polysaccharide and mucopolysaccharide hydrolysing enzymes* | | |
| β-Galactosidase | Galactosides | Animals, plants, protists |
| α-Glucosidase | Glycogen | Animals |
| α-Mannosidase | Mannosides | Animals |
| β-Glucuronidase | Polysaccharides and mucopolysaccharides | Animals; plants |
| Lysozyme | Bacterial cell walls and mucopolysaccharides | Kidney |
| Hyaluronidase | Hyaluronic acids; chondroitin sulphates | Liver |
| Arylsulphatase | Organic sulphates | Liver; plants |
| *Proteases* | | |
| Cathepsin(s) | Proteins | Animals |
| Collagenase | Collagen | Bone |
| Peptidases | Peptides | Animals; plants; protists |
| *Lipid degrading enzymes* | | |
| Esterase(s) | Fatty acid esters | Animals; plants; protists |
| Phospholipase(s) | Phospholipids | Animals; plants? |

found to be active in the supernatant fluid fractions of the particle suspensions. It became apparent to the Belgian group that they were dealing with an organelle possessing a wealth of potential for metabolic disorganization since enzymes capable of degrading almost all the important macromolecules of the cell were separated from cytoplasmic substrates by a delicate *single lipoprotein membrane*. It was difficult at that time to assign any definite function to these organelles since intracellular digestion was considered, with the exceptions of leucocytes and macrophages, an exclusive attribute of primitive organisms. However, the potentially lytic nature of such an arrangement did not escape attention and, appropriately, the word *lysosome* was proposed for the membrane-bounded particle containing the five hydrolases: acid phosphatase, acid ribonuclease, acid deoxyribonuclease, β-glucuronidase and cathepsin. It was some time later that the importance of lysosomes in cell injury, disease and death was realized. The early concept of the lysosome is illustrated in Fig. 1.1.

The list of known lysosome enzymes continues to lengthen rapidly and at least forty have been located in a variety of tissue types up to the present. Some of the enzymes found in lysosomes of a number of tissues are shown in Table 1.1, many others have been located at such lysosomal sites in animal tissues and extensive lists of the enzymes and their properties have been compiled by Tappel (1969) and by Barrett (1972). It is obvious even from the limited number of enzymes in this table that lysosomes have the potential to degrade virtually every macromolecule of biological origin. Biochemical studies and E.M. cytochemistry show that the lysosome population is heterogeneous with respect to enzyme content and a single lysosome need not contain the full spectrum of enzymes. Many of the enzymes also exist as several molecular forms which may have distinctive locations in various parts of the cell, tissue or organism, and these may vary with the physiological state of the tissues. Thus, four molecular forms of acid phosphatase capable of hydrolysing nitrophenol phosphate have been located at four specific sites, namely, the large granule, lysosome, microsome, and supernatant fractions of guinea pig and mouse liver. A similar situation has been found in potato sprout homogenates where 'Heavy' and 'Light' lysosome fractions contained acid phosphatases differing in molecular weight (Fig. 1.2). Thus, in recapitulation, *lysosomes may be variable in respect of enzyme content* and *acid hydrolases may frequently occur at extralysosomal sites*.

Immediately following the publication of de Duve's findings Straus, working independently in America, showed by an important series of experiments that kidney lysosomes in rats were able to accumulate injected foreign proteins, such as egg white and horseradish peroxidase, and act in detoxification (Straus, 1967, reviews this work), thus indicating a connection between lysosomal digestion and the well-known process of endocytosis. Extensive progress followed these observations and a number of research teams were able to link-up lysosome function with well-documented observations from the older literature. It was a time of important discoveries and it may be difficult for contemporary students, reared on a diet of cell compartmentation, to appreciate the excitement engendered by this work, but it may be worthwhile observing that this was an important period which moulded our current ideas of the nature of sub-cellular organization which up until that time extended only a little beyond the concept of the cell as a bag of enzymes.

Although the lysosome *per se* is a relatively new organelle compared to the nucleus, chloroplast and Golgi apparatus, a mass of information has accumulated concerning its structure, heterogeneity and functions in both normal and diseased tissues through the combined use of biochemical, cytological and ultrastructural techniques. However, the history of the lysosome extends retrospectively into the last century, and the elegant review of de Duve (1969) relates the work of Metchnikoff on intracellular

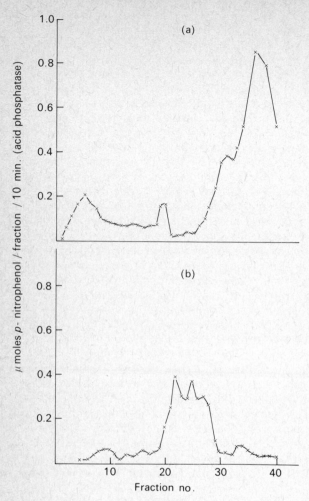

**Fig. 1.2** Evidence for the different sub-cellular locations of acid phosphatase in potato shoots. Gel filtration elution patterns of molecular forms of acid phosphatase capable of hydrolysing *p*-nitrophenyl phosphate from (*a*) heavy lysosome fraction and (*b*) light lysosome fraction. (From Pitt, D. and Galpin, M. (1973) *Planta (Berl.)*, 109, 233–58, © Springer-Verlag, 1973.)

granules and of Ehrlich on vital dye uptake to the modern concept of the lysosome not as an isolated body but as part of a complex *dynamic system* involved in mechanisms of cellular absorption, digestion, storage, secretion and autophagy.

## 1.2    Methods Used in the Study of Lysosomes

Before moving to the next two important considerations which concern
the origin and fate of lysosomes and the nature of the lysosome membrane
I feel it is necessary to mention briefly some of the basic techniques which
provided the experimental basis and impetus for the development of the
lysosome concept. I believe such a treatment may simultaneously convey
some understanding of the physical and biological properties of the
organelle.

### 1.2.1    Separation of lysosomes
Lysosomes are extremely delicate organelles which are readily damaged by
the vigorous techniques of conventional 'solution biochemistry' that
prevailed in almost all the pioneering biochemical work on metabolic pathways. It was the development of milder techniques for the disruption of
cells coupled with the use of isotonic media that led to the detection of
organelles such as lysosomes and peroxisomes in animal tissues. The
extreme fragility of the lysosomes *in vitro* has hindered their isolation
from many plant materials which normally require for the separation of
cells and the breaking of cell walls, high shearing forces that also destroy
the integrity of lysosomes. An homogenate containing a high proportion
of intact organelles is layered over a linear gradient of an inert solute,
often sucrose, contained in a plastic centrifuge tube which is then subjected to centrifugation in a preparative ultracentrifuge operating near
4° C. During centrifugation the various organelles in the original homogenate move different distances according to their sedimentation constants
which are related to particle size, shape and density. The tubes may then
be drained through a puncture in the base, and the contents collected as
fractions of known volume by means of a fraction-collecting device.
Such an isolation procedure, outlined in Fig. 1.3, has now been highly
developed to permit the simultaneous large-scale separation of lysosomes
and other organelles by zonal centrifugation or by continuous fractionation using special continuous-feed centrifuges. Various refinements have
also been introduced into gradient composition since sucrose enters particular organelles to differing extents thereby influencing their densities
and hence the subsequent fractionation procedures. Once the fractionation
has been completed the organelles within the isolated fractions may be
characterized by biochemical and ultrastructural methods. A critical
account of isolation and fractionation techniques with particular reference
to lysosomes is presented by de Duve (1967).

### 1.2.2    Cytochemistry of lysosomes
Whereas the above isolation procedures are invaluable in the study of cell
compartments it must be remembered that observations are being made *in
vitro* under conditions which are artificial and on cell particles which may

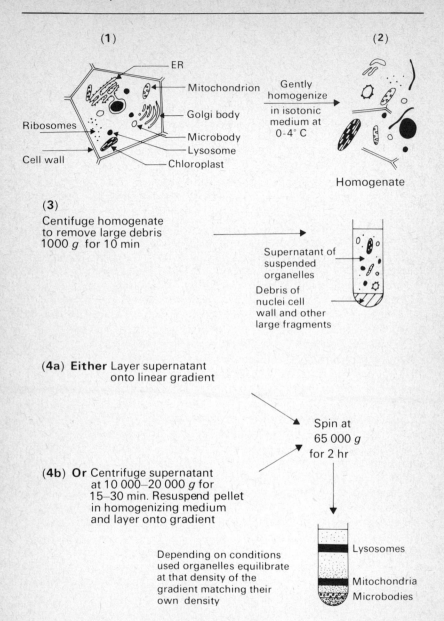

**(1)**

ER

Mitochondrion

Golgi body

Ribosomes

Microbody

Cell wall

Lysosome

Chloroplast

Gently
homogenize

in isotonic
medium at
0-4° C

**(2)**

Homogenate

**(3)**

Centifuge homogenate
to remove large debris
1000 *g* for 10 min

Supernatant of
suspended
organelles

Debris of
nuclei cell
wall and other
large fragments

**(4a) Either** Layer supernatant
onto linear gradient

Spin at
65 000 *g*
for 2 hr

**(4b) Or** Centrifuge supernatant
at 10 000–20 000 *g* for
15–30 min. Resuspend pellet
in homogenizing medium
and layer onto gradient

Depending on conditions
used organelles equilibrate
at that density of the
gradient matching their
own density

Lysosomes

Mitochondria

Microbodies

**Fig. 1.3** Outline of a cell fractionation procedure for separating lysosomes from other organelles in a tissue homogenate using density gradient ultracentrifugation.

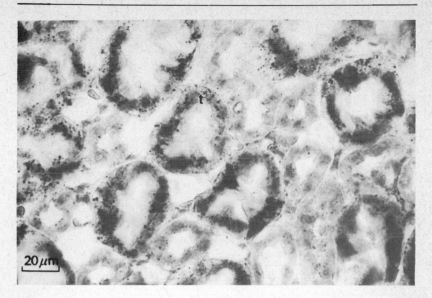

**Fig. 1.4** Frozen section of formol-calcium fixed rat kidney incubated for 20 min. at 37° C in the β-glycerophosphate medium of Gomori. The reaction product, phosphate, is converted at the site of its production to black lead sulphide, the deposits of which are indicative of acid phosphatase activity. Much of the activity is confined to the lysosomes, particularly of the convoluted tubules (t).

have suffered gross damage. It is also impossible to study integrated cell processes in the test-tube. Consequently the development of cytochemical staining procedures for lysosome enzymes was a vital step towards an understanding of the nature of the lysosome system. It is particularly fortunate that the hydrolases of lysosomes are amongst the few enzymes that are amenable to visualization by cytochemical methods and also withstand some of the fixation procedures necessary for study in both the light and electron microscopes. The phosphatase method, described by Gomori well before lysosomes were discovered, has been particularly useful and is based on the principle that inorganic phosphate split from certain phosphate esters by acid phosphatase, in correctly prepared tissue sections, reacts with lead ions at the sites of enzyme activity yielding lead phosphate which is rendered visible for light microscopy by reaction with ammonium sulphide to give a brown stain. Lysosomes visualized in the light microscope by the Gomori technique are illustrated in Figs. 1.4 and 1.5. The technique is of great value in electron microscope studies when the lead phosphate itself is electron dense (Fig. 1.6). Special modifications of this classic procedure have been used to detect other lysosome enzymes including deoxyribonuclease and ribonuclease. Certain drawbacks of the

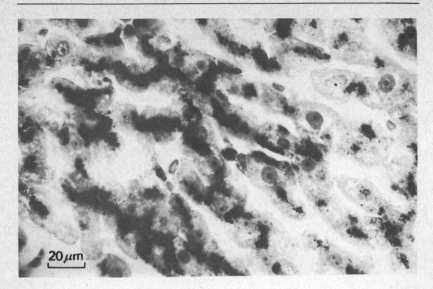

**Fig. 1.5** Rat-liver section stained for acid phosphatase activity as in Fig. 1.4, but with 30 min. incubation, indicating lysosomal sites of activity.

method, revealed by critical work, include a high degree of non-specific staining in some tissues due to reaction with lead ions of endogenous tissue phosphates which may be embarrassingly high in some fungi and higher plants. However, great impetus was imparted to the study of lysosomes when it was found that certain phosphate esters of substituted naphthols were hydrolysed by acid phosphatase in tissue sections, and that the liberated naphthol could be reacted with various diazonium compounds to give coloured products at the putative enzyme site. It is now possible, through the use of appropriate substrates, to extend the azo dye technique to a number of other lysosome enzymes including β-glucuronidase, β-galactosidase and N-acetyl-β-glucosaminidase. The various modifications of this technique show a good degree of specificity and produce highly substantive and insoluble end-products, many of which may be detected by high resolution electron microscopy. The latter has revealed important *extralysosomal sites* for a number of acid hydrolases which were formerly believed to have exclusive lysosome locations. Sites of acid phosphatase activity in the ER (endoplasmic reticulum), Golgi apparatus and ribosomes of animal tissues have permitted the development of important concepts on the origin of lysosomes which we consider in the next section. It is also recognized that similar extralysosomal locations of acid phosphatase exist in protists and higher plants and in the latter additional sites of hydrolases can be demonstrated in the plasmodesmata and the cell walls. For those

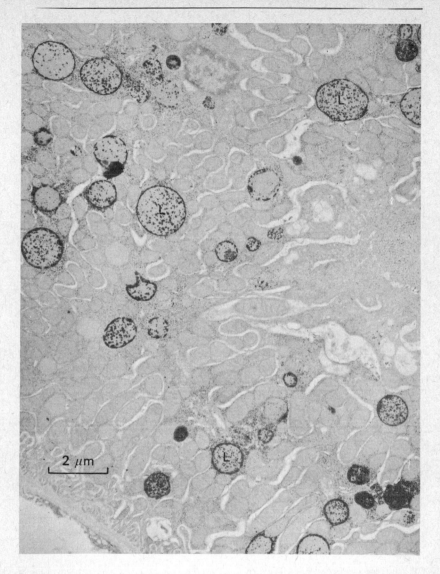

**Fig. 1.6** Electron micrograph of a thin section from glutaraldehyde/cacodylate fixed rat kidney showing localization of acid phosphatase by the Gomori procedure. Electron dense (black) deposits show the sites of acid phosphatase activity mainly at the membranes of and within lysosomes (L). (From Novikoff, A. B. (1963) in: *Ciba Foundation Symposium on Lysosomes*, J. & A. Churchill, Ltd. By permission of the publishers. Original print supplied by Professor A. B. Novikoff.)

requiring a critical coverage of the histochemistry of lysosomes the review of Gahan (1967) should be consulted.

### 1.2.3 Markers for endocytic vacuoles

An early stimulus to the work on lysosomes came from Straus's observations that when foreign proteins were injected into mammals they were ingested in both liquid (pinocytosis) or solid (phagocytosis) forms by cells in the generalized process of *endocytosis*. Thus, foreign proteins, colloids of heavy metals, carbon and other particulate materials which are detectable by light or electron microscopy can be seen accumulating especially in liver and kidney cells. The fate of such materials has been carefully followed, in particular that of injected peroxidase which is conveniently subject to cytochemical detection by the extremely sensitive benzidine reaction. Exogenous protein accumulating at sub-cellular sites, i.e. peroxidase, stains blue, and is initially confined to the phagosomes but eventually finds its way into secondary lysosomes. These important experiments demonstrated a key role of lysosomes in intracellular digestion. This process can now be followed by electron microscopy since the marker enzymes for phagosomes (peroxidase) and lysosomes (acid phosphatase) can generate electron-dense reaction products under appropriate conditions. We shall refer again to this valuable technique in the section on intracellular digestion (Chapter 3, p. 58).

A further related observation showed that when the detergent Triton WR-1339 was injected into rats it accumulated in the liver lysosomes which then increased in size and decreased in density thus allowing a cleaner separation from other organelles of similar original density. This technique has been of value in biochemical work when 'pure' lysosome fractions were required, but the lysosomes so obtained may show some abnormal properties.

## 1.3 The Origin and Fate of Lysosomes

Techniques of electron microscopy and biochemistry reveal lysosomes as heterogeneous organelles differing greatly in size, density and structure. These diverse physical characteristics coupled with divergent biological properties have made it impossible to separate completely the various components of the lysosome system in pure states in spite of extensive efforts. Consequently most of our knowledge derives from work with partially purified fractions from specific types of tissues, particularly rat liver and kidney which prove exceptionally amenable to study. It soon became apparent that the biochemical differences associated with lysosomes were related to the dynamic nature and range of functions of this system in the life of the cell. Separation of the lysosome system into a number of its components is now possible by means of sophisticated centrifugation techniques and I have already mentioned that some enzymes and isoenzymes may have different locations in the various parts of the system. However, care is necessary in the interpretation of such results since the fragile nature of lysosomes makes it difficult to exclude the possibility of false locations following both organelle damage and re-adsorption of the enzymes onto various types of cell debris. In spite of these morphological and biochemical heterogeneities which have complicated the studies on lysosomes compared to other organelles it is generally considered that sufficient similarities exist between lysosomes from various tissues to justify their inclusion into a single group of particles rather than to subdivide them into numerous separate groups of organelles.

Although it is possible using cytochemical methods to demonstrate dual localizations of enzymes within a particular lysosome it is not yet known if all the lysosomes of a cell contain every lysosomal enzyme or if specific components of the system contain only some of the spectrum of enzymes detectable in isolated populations of lysosomes by biochemical means. The results from the combined use of techniques of electron microscope cytochemistry and of biochemistry allow the assumption that *populations* of lysosomes in all the tissues examined are involved in and could potentially complete the hydrolysis of all the macromolecules found in normal cells. Under such circumstances the material to be digested must gain access to the interior of the lysosome without the liberation of hydrolases to the cytoplasm. Detailed observations on different cell types, particularly tissue culture cells, reveal some diversity in the process but we are now aware of a general pattern of lysosome function due to the efforts of Novikoff and others.

Such studies suggest the existence of two main sub-cellular routes in the production of *primary* lysosomes. One school of thought supports the view that primary lysosomes arise in the same manner as secretory vesicles of certain secretory cells and in particular follow the classical pattern demonstrated in Palade's laboratory for the secretion of exocrine products of the pancreas. Those proteins destined for lysosomal sites are synthesized on the ribosomes attached to the limiting membrane of the rough endoplasmic reticulum (RER). They eventually enter the cisternae of the RER and are transferred to the Golgi apparatus which they enter as small vesicles via the convex (forming) face. During passage through the Golgi complex they are packaged into lysosomes within the condensing vacuoles of the apparatus and eventually leave at the concave (emitting or secreting) face. Opinion varies whether the emergent primary lysosomes do or do not contain *active* hydrolases. An alternative pathway has been proposed by Novikoff and co-workers who have developed the concept of the GERL (= *G*olgi-associated *E*ndoplasmic *R*eticulum from which *L*ysosomes arise) in which this specialized portion of the smooth endoplasmic reticulum (SER), distinct from but adjacent to the *emitting* face of the Golgi body, is involved in the genesis of primary lysosomes. It is envisaged that synthesis of lysosome enzymes occurs at the RER and are then subsequently transported to the region of the GERL where the acid phosphatase-positive portion of the ER is budded-off as primary lysosomes. In this scheme the enzyme proteins of the primary lysosomes by-pass the Golgi apparatus and the membranes are not differentiated by passage through the Golgi complex. Whatever its derivation the primary lysosome is thought by many to be of a somewhat evanescent nature and is not itself involved in digestive events. Primary lysosomes are believed to merge fairly soon after their formation with phagosomes, pinocytic or autophagic vacuoles to form the main digestive unit known as the *secondary lysosome*. Presumably any inactive hydrolase within the primary lysosome becomes activated at this stage. Repeated fusions may occur between phagosomes, primary lysosomes and secondary lysosomes which may result in the formation of a multivesicular body which also possesses digestive capacity and is classified as a secondary lysosome. There are no biochemical or cytochemical doubts that the newly-formed secondary lysosomes contain very high acid hydrolase activity and it is within these structures that sequestered materials are partially or completely digested with undegradable materials remaining as residues. Secondary lysosomes that are involved in repeated fusions with phagosomes may be 'topped-up' with enzymes from primary lysosomes. Eventually they become depleted of hydrolase activity and are designated *residual bodies*. The latter are not generally extruded to the exterior of the cell, but normally remain indefinitely within the cells. Extensive observations reveal that a further component of the lysosome system, the *autophagic body*, may be formed from the SER either with or without the involvement of the Golgi apparatus. The formation, nature and function of these entities is discussed in a later section (Chapter 3, p. 64).

Many ultrastructural observations testify to the variations in the processes of lysosome formation and the interrelationships of the several components of the system. The reported diversities probably relate to the variation in techniques as well as to the nature and functions of particular

---

Fig. 1.7    Suggested origins, interrelationships and roles of lysosomes. Shaded areas indicate hydrolase activity. The scheme is based largely on the proposals of Novikoff and co-workers and the diagram is essentially that in *Cells and Organelles* by Novikoff and Holtzman (1970), which is reproduced by permission of Holt, Rinehart and Winston, Inc.

In *animal cells* primary lysosomes arise from the Golgi apparatus via Route I and as vesicles derived from the ER (Route II(i)) or from the GERL (Route II(ii)). Autophagic bodies (secondary lysosomes) are thought to originate through Route III. Route IV also seems a functional primary lysosome source.

In *plant cells* Routes I and II are thought to predominate, but evidence for the existence of the GERL is slight. Although autophagic bodies are common in plants their origin(s) is obscure.

In *Protista* evidence from the Protozoa indicates the likely existence of Routes I, II, III and IV. In fungi a special pathway (Route V) seems important in extracellular secretion, and may extend to other groups.    ▶

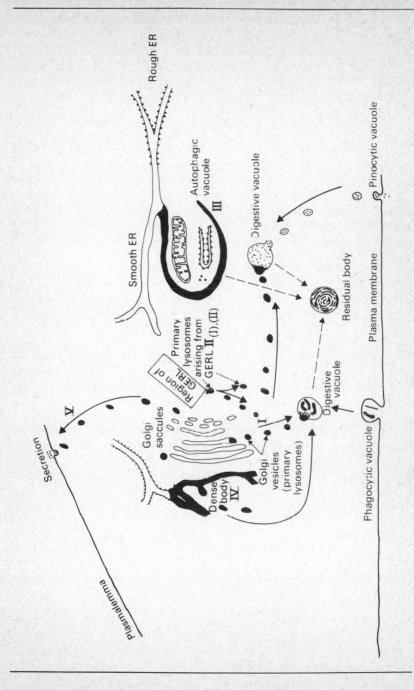

Rough ER

Smooth ER

Autophagic vacuole

III

Digestive vacuole

Pinocytic vacuole

Residual body

Plasma membrane

Primary lysosomes arising from GERL II(I),(II)

Region of GERL

Golgi saccules

I

Digestive vacuole

Golgi vesicles (primary lysosomes)

Phagocytic vacuole

V

Secretion

Dense body IV

plasmalemma

cell types, but a generalized scheme based heavily on that presented by Novikoff is illustrated in Fig. 1.7. Possible involvement of the Golgi apparatus in these processes is fully in line with current ideas on the function of this organelle as reviewed by Dauwalder *et al.* (1972). Recent improved techniques for isolating the Golgi complex should help towards providing some clarification of the origin and interrelationships of the various components of the lysosomal system.

A lively literary debate is in progress concerning the origins of lysosomes in plants. From earlier work clear indications emerged that these may arise from the ER, but the possible involvement of the Golgi apparatus was not excluded. Current cytochemical work being done in France (Coulomb and Coulomb, 1971; Marty and Buvat, 1972) reveals acid phosphatase at sites in both the ER and the Golgi complex from which dense bodies (primary lysosomes) emerge and fuse with autophagic bodies in a way extraordinarily reminiscent of the manner in which secretory proteins of the pancreas are packaged. However, other workers have frequently failed to locate acid phosphatase, or any other acid hydrolases, in the Golgi complex of plant cells. Thus, although the majority of evidence from plant studies regards provacuoles and vacuoles as lysosomes derived from the ER, I believe that insufficient observations have yet been published to allow a clear assessment of lysosome origins to be made. It should be of some concern to plant biologists that so few biochemical studies have been made on this topic and the more frequent ultrastructural cytochemical studies using the rather capricious Gomori procedure have rarely been complemented by experimental biochemistry with the result that dynamic conclusions regarding the lysosome system have been drawn, all too frequently, from the valuable yet static images obtained from the electron microscope.

The origins of lysosomes in the Protista are also unclear. Early work on *Tetrahymena pyriformis* provided evidence in support of the GERL concept, but recently exciting work on fungi by Grove *et al.* (1968) has provided striking evidence for formation via the Golgi apparatus of what might be considered lysosomes. We shall refer to this work on other occasions in this book. The notion also arose from early work on *Euglena gracilis*, and now extends to some higher organisms, that primary lysosomes arise as *dilations* of the SER and of the Golgi apparatus. Current work on *Paramecium* also confirms a similar multiple origin of lysosomes.

I hope that this rather brief discussion of the various proposals on the origins and interrelationships of the components of the lysosome system will form sufficient basis for the more detailed treatments of specific aspects which recur throughout the remainder of this book. For those seeking more extensive coverage on this fascinating facet of cell biology the stimulating recent articles by Novikoff (1973) and Novikoff *et al.* (1971) are recommended since they provide a wealth of both established and speculative ideas.

## 1.4 The Lysosome Membrane

The fundamental importance of the lysosome membrane as part of the lysosome concept was recognized early in the studies on this organelle. Thus, 'latency' of lysosome enzymes *in vitro* was dependent upon the integrity of the membrane with 'activation' of these enzymes as a feature associated with treatments causing membrane damage. Initially these treatments included mechanical disruption, freezing and thawing, osmotic shock and use of phospholipases, proteases and detergents. The literature abounds with information concerning other factors which cause instability of lysosome membranes. Included in these *labilizers* are vitamins A and D, some steroids, bacterial toxins and ultraviolet irradiation; some of which we shall examine again in later chapters. In addition there exists a small group of compounds which have a *stabilizing* action on lysosomes, probably through their effects upon the lipid components of the lysosome membrane. Hydrocortisone is particularly important in this respect and has a striking effect on inhibiting the hydrolase release mediated by ultraviolet irradiation and also markedly reduces the capacity of lysosomes to digest foreign matter, presumably through antagonizing membrane fusion. Because of the extreme relevance of these observations to physiology and medicine, the modes of action of both labilizers and stabilizers is a field of intense research activity.

The first observations on the activation of lysosome enzymes from isolated lysosomes were explained in terms of a release of enzymes in solution from a membranous sac thus permitting access to substrates in the suspending medium. It was, however, difficult to explain how a mixed solution of lytic enzymes exists within a lysosome without the enzymes themselves entering into mutual digestion or degradation of the lysosome membrane. Nowadays the original idea has been modified since techniques of biochemistry and cytochemistry show that whereas some enzymes from rat liver lysosomes, e.g. $\beta$-galactosidase and $\alpha$-mannosidase are readily and completely released following conventional treatments others such as $\beta$-glucosidase remain firmly bound to the membrane. It is also known that all lysosome enzymes are not activated at the same rate. The exact means by which enzymes are bound is not clear, but Koenig has proposed, following extensive work, that the lysosome may be considered as a membrane-bounded polyanionic lipoprotein granule in which the enzymes are held in an inert condition through electrostatic binding to the acidic groups of the lipoprotein matrix. In support of this hypothesis are the observations that extremes of pH and certain critical levels of cations, such as $Ag^+$, $Hg^{2+}$ and $Cu^{2+}$, bring about release of lysosome enzymes without any discernible structural damage to the lysosome. Furthermore, pertinent observations suggest that the lysosome membrane may require gross biochemical modification before certain of the enzymes are released. Thus, experiments with phospholipase C brought about the simultaneous release

17

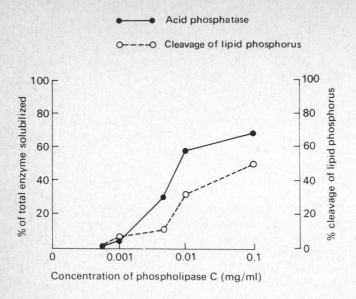

Fig. 1.8   Effects of phospholipase C on the solubilization of acid phosphatase and cleavage of lipid phosphorus of rat-liver lysosomes. (Data redrawn from Koenig, H. (1969) in: *Lysosomes in Biology and Pathology*, eds. J. T. Dingle and H. B. Fell, vol. 2, ch. 6. By courtesy of North-Holland Publishing Co.)

of several latent hydrolases of lysosomes with the coincident cleavage of phospholipids (Fig. 1.8). Recent work by the author on lysosomes from potato shoots showed that a considerable proportion of the acid phosphatase was so tightly bound to the membrane as to resist the solubilizing effects of detergents, ultrasonication and lipid solvents, with release only being significantly achieved by extremes of pH (Fig. 1.9) or by digestion with purified phospholipase A or snake venom (Fig. 1.10).

A further point of interest is the finding that the activity of some lysosome enzymes, e.g. ribonuclease, is dependent upon the existence of sulphydryl groups within the molecule. Oxidation of these to disulphide bonds results in reversible inactivation of the enzymes. It may be that some lysosomal hydrolases could be linked to the membrane through disulphide bonds and will be inactive until re-activated via an oxidation-reduction system. Some evidence, therefore, now favours the view that at least a component of the lysosome enzyme complement is structurally associated with the membrane, but it cannot be excluded that some of the enzymes also occur in a soluble phase within the lysosome. It is also possible that the location and state of enzymes within the lysosomes may

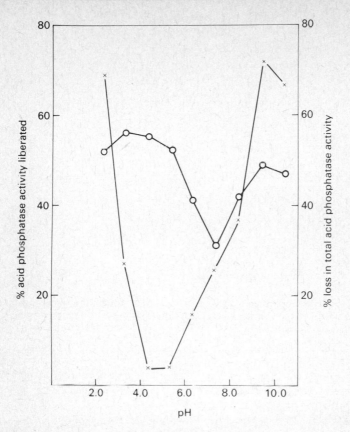

Fig. 1.9  Effects of pH upon liberation of acid phosphatase from isolated potato shoot lysosomes; x ————— x enzyme liberated; o ————— o loss of activity. (From Pitt, D. and Galpin, M. (1973) *Planta (Berl.)*, **109**, 233–58. © Springer-Verlag, 1973.)

be influenced by and dependent upon the physiological state of the lysosome and the particular tissue in question.

Other major problems exist relating to the lysosome membrane, and questions concerning membrane structure and fusion with other components of the membrane system remain largely unanswered. Virtually no information is available about the possible passage of substrates of internal origin into (apart from autophagy) and the transport of products of digestion out of lysosomes, although the presence of adenosine triphosphatase (ATP-ase) in the membrane suggests the presence of a transport system which is ATP-dependent. Indeed, it has been shown that ATP-stimulated secretion of β-glucuronidase and ribonuclease occurs from leucocytes in

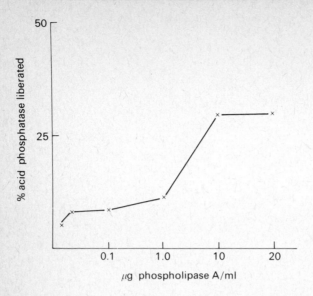

Fig. 1.10    Liberation by various concentrations of purified phospholipase A of lysosomal acid phosphatase from a crude lysosome fraction from potato shoots. (From Pitt, D. and Galpin, M. (1973) *Planta* (*Berl.*), 109, 233—58. © Springer-Verlag, 1973.)

the presence of calcium ions and so suggests similarities with well-established membrane transport phenomena.

From the scheme outlined in Fig. 1.7 it is apparent that controlled mechanisms of membrane fusion are of fundamental importance in lysosome physiology. Secretion or extrusion of materials from lysosomes to the outside of the cell involves fusion between the plasma membrane and the lysosome membrane; whilst intracellular digestion depends upon fusion between endocytic vacuoles, the membranes of which being largely derived from the plasma membrane, and the primary lysosomes. The production of autophagic vacuoles is another key process which depends on membrane fusion. It is obvious that the ordered functioning of the cell requires some subtle mechanism which prevents random fusion of lysosomes with other organelles such as mitochondria, chloroplasts and nuclei. Likewise, some mechanism must prevail which prevents random 'gulping' of the cytoplasm by uncontrolled autophagy. Manipulation of the fusion process has been possible by the use of retinol (vitamin A alcohol), which acts as a surfactant, such that primary lysosomes (or Golgi-derived vesicles) fuse directly with the plasma membrane and achieve extracellular release of the enzymes *in vivo*. In such experimentally-induced situations there are few cytopathic effects, and of particular relevance is the observation that

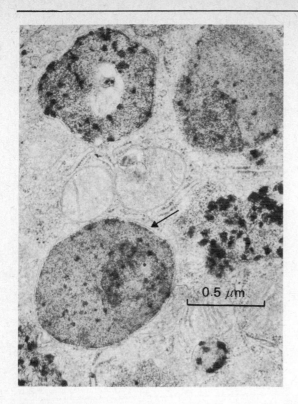

Fig. 1.11   Electron micrograph of a thin section of the cytoplasm of a red-pulp macrophage of mouse spleen. Electron-dense lead deposits reveal the presence of acid phosphatase in membrane-limited inclusion bodies (lysosomes). Note the clear image of the single membrane (arrow). (From Daems, W. Th. *et al.* (1969), in: *Lysosomes in Biology and Pathology*, eds. J. T. Dingle and H. B. Fell, vol. 1, ch. 3. By courtesy of North-Holland Publishing Co. Original print supplied by Professor W. Th. Daems.)

the promotion of extracellular release is not accompanied by random fusion of the potentially destructive primary lysosomes with other organelles. These important findings led Dingle to suggest that the fusion of lysosome membranes with other parts of the membrane system is controlled by surface phenomena such that vesicles will only fuse if they each possess high surface tension. As a consequence he predicted that the membranes of the Golgi-derived vesicles have a high surface tension and should be able to fuse only with the plasma membrane or derivatives of it. It is difficult to provide experimental data supporting this hypothesis and the idea has been disputed, particularly by J. A. Lucy, who proposed that the lipid component of membranes exists as spherical micelles embedded in the membrane and that the pre-requisite for the fusion of two membranes is that both should possess a high proportion of their phospholipid in a micellar configuration. Fusion would not occur, therefore, if either or both were low in micelles and approached the bimolecular leaflet configuration.

Both of these notions are speculative and neither may apply since our ideas on the structure and 'flow' of lysosome membranes are rudimentary as are those relating to biological membranes in general. For those wishing to pursue the matter a recent essay tracing the development of ideas on membrane structure has appeared (Finean, 1972), and Lucy (1969) has reviewed our knowledge of lysosome membranes. It is clear though that an understanding of the structure and fusion of lysosome membranes is vital to future developments in the field of lysosome research.

## Suggestions for further reading

BARRETT, A. J. (1972) Lysosomal enzymes, in *Lysosomes — a laboratory handbook*, ch. 2, pp. 46—135, ed. J. T. Dingle. North-Holland Publishing Co., Amsterdam and London.

COULOMB, P. and COULOMB, C. (1971) Phytolysosomes: Mise en évidence de relationships entre le réticulum endoplasmique, les dictyosomes et les Phytolysosomes dans le méristème radiculaire de la Courge (*Cucurbita pepo*, L. Cucurbitaceae). *C.R. Hebd. Seances Acad. Sci. Ser. D. Sci. Natur.*, 272, 1634—7.

DAUWALDER, M., WHALEY, W. G. and KEPHART, J. E. (1972) Functional aspects of the Golgi apparatus, *Sub-cell. Biochem.*, 1, 225—75.

de DUVE, C. (1967) General principles, in *Enzyme Cytology*, ch. 1, pp. 1—26, ed. D. B. Roodyn. Academic Press, London and New York.

de DUVE, C. (1969) The lysosome in retrospect, in *Frontiers of Biology*, 14A. *Lysosomes in Biology and Pathology*, vol. 1, ch. 1, pp. 1—40, eds. J. T. Dingle and H. B. Fell. North-Holland Publishing Co., Amsterdam and London.

de DUVE, PRESSMAN, B. C., GIANETTO, R., WATTIAUX, R. and APPELMANS, F. (1955) Tissue fractionation studies. 6. Intracellular distribution patterns of enzymes in rat-liver tissue, *Biochem. J.*, 60, 604—17.

FINEAN, J. B. (1972) The development of ideas on membrane structure, *Sub-cell. Biochem.*, 1, 363—73.

GAHAN, P. B. (1967) Histochemistry of lysosomes, *Int. Rev. Cytol.*, 21, 1—63.

KOENIG, H. (1969) Lysosomes in the nervous system, in *Frontiers of Biology*, 14B. *Lysosomes in Biology and Pathology*, vol. 2, ch. 6, pp. 109—62, eds. J. T. Dingle and H. B. Fell. North-Holland Publishing Co., Amsterdam and London.

LUCY, J. A. (1969) Lysosomal membranes, in *Frontiers of Biology*, 14B. *Lysosomes in Biology and Pathology*, vol. 2, ch. 6, pp. 313—41, eds. J. T. Dingle and H. B. Fell. North-Holland Publishing Co., Amsterdam and London.

MARTY, F. and BUVAT, R. (1972) Distributions des activités phosphatasique acides au cours du processes d'autophagie cellulaire dans les cellules du meristeme radiculaire d'*Euphorbia characias* L., *C.R. Acad. Sci.* (Paris), 274, 206—9.

NOVIKOFF, A. B. (1973) Lysosomes: a personal account, in *Lysosomes and Storage Diseases*, ch. 1, pp. 1—41, eds. H. G. Hers and F. van Hoof. New York, Academic Press.

NOVIKOFF, A. B., ESSNER, E. and QUINTANA, N. (1964) Golgi apparatus and lysosomes, *Fedn. Proc. Fedn. Amer. Socs. exp. Biol.*, 23, 1010—22.

NOVIKOFF, P. M., NOVIKOFF, A. B., QUINTANA, N. and HAUN, J—J. (1971) Golgi apparatus, GERL and lysosomes of neurons in rat dorsal root ganglia studied by thick and thin section cytochemistry, *J. Cell Biol.*, 50, 859—66.

PITT, D. and GALPIN, M. (1973) Isolation and properties of lysosomes from dark-grown potato shoots, *Planta (Berl.)*, 109, 233—58.

STRAUS, W. (1967) Lysosomes, phagosomes and related particles, in *Enzyme Cytology*, ch. 5, pp. 239—319, ed. D. B. Roodyn. Academic Press, London and New York.

TAPPEL, A. L. (1969) Lysosomal enzymes and other components, in *Frontiers of Biology*, 14B. *Lysosomes in Biology and Pathology*, vol. 2, ch. 9, pp. 207—44, eds. J. T. Dingle and H. B. Fell. North-Holland Publishing Co., Amsterdam and New York.

WATTIAUX, R. (1969) Biochemistry of lysosomes, in *Handbook of Molecular Cytology*, pp. 1159—78, ed. A. Lima-de-Faria, North-Holland Publishing Co., Amsterdam and London.

23

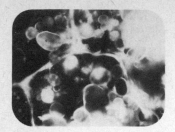

# Chapter 2

# Occurrence of Lysosomes

Lysosomes were initially defined using biochemical criteria because most of the early work on these organelles was primarily of a biochemical nature. This definition considered *lysosomes as a group of sub-cellular particles bounded by a single limiting membrane and containing a number of acid hydrolases characterized by pH optima around pH 5.0.* No internal structure was associated with the lysosomes, and release of the hydrolases accompanied by 'activation' was considered a result of membrane disruption. This definition was based largely on initial work with rat liver and kidney and a few other mammalian tissues. Slowly it became apparent, however, that lysosomes were a variable group of particles differing in their distribution within tissues and organs. Work was complicated by the heterogeneity of tissues which was responsible for the inability to obtain lysosomes from any specific cell type of a particular organ. Many tissues were not suited to those techniques so readily applicable to rat liver and kidney and it was not long before the value of cytological and cyto-chemical techniques was realized. Using such methods the estimation of the number and distribution of lysosomes in different cell types and in various parts of organs could be easily and quickly obtained. Gradually, generally accepted criteria for the histochemical identification of lyso-somes emerged whereby a *lysosomal location depended on demonstrating the presence of two or more acid hydrolases at a particulate site that was limited by a single membrane.* It was also possible to demonstrate the phenomenon of structure-linked latency in tissue sections by appropriate use of the Gomori technique. Thus, sections of unfixed rat liver, obtained by sectioning fresh frozen tissues, required 20 min. incubation before reaction product could be detected, whilst formalin-fixed sections produced visible end-product after 10 min. incubation. The longer incubation period required in the case of unfixed tissues was considered to arise through the impermeability of the intact membranes to molecules of enzymes and substrates, with delayed activation being mediated by the acid pH of the medium and other incubation conditions which caused eventual membrane damage. Formalin fixation was believed to affect the structure of the membrane to allow immediate access of substrate to enzyme. Structure-linked latency at the histochemical level has since been demonstrated in many more animal and some plant and fungus tissues. These observations coupled with the impressive achievements of histochemists in locating hydrolases by means of electron microscopy led to rapid progress in the study of lysosomes which was becoming restricted by the limitations imposed by accepting the original biochemical definition.

Over recent years other properties of lysosomes have been recognized using the elegant and valuable peroxidase technique which has now been modified and extended to demonstrate micropinocytic bodies by electron microscopy. Other useful electron-dense markers including colloidal metals, carbon and thorium dioxide, which are taken up into phagosomes, have been used as lysosome markers at the level of ultrastructure. It has also proved possible to use non-enzymic structural components as markers of lysosomes. Thus, since the membrane of lysosomes contains appreciable amounts of phospholipid which is detectable at both the light microscope and electron microscope levels by specific staining procedures, this component has occasionally served as a lysosome marker but some difficulties are experienced through lack of specificity since most of the membrane system of the cell is rich in phospholipids. Lysosomes also exhibit metachromasia with toluidine blue suggesting that they may contain acid mucopolysaccharides. This characteristic has been used extensively in some work on chick fibroblasts and rat ovuli. In addition lysosomes contain other types of polysaccharide material which react with periodic-acid-Schiff stain and provide a further means of identifying these structures, particularly in liver parenchyma cells, brain tissues and chick fibroblasts. Lysosomes also have the interesting capacity to accumulate vital dyes, some of which may have fluorescent properties that make for spectacular viewing by light microscopy. Under appropriate conditions acridine orange, euchrysine and the carcinogenic hydrocarbon 3,4-benzpyrene, accumulate and fluoresce brilliantly at lysosome sites when illuminated by blue light (see Appendix, p. 150). This method has proved particularly useful for marking lysosomes in living tissues and permits observation of their behaviour during tissue damage, infection and other cytopathic conditions. The fluorescence microscope has also been valuable in the localization of lysosomes using fluorescent labelled proteins. By this immunological technique tissue sections are treated with fluorescein-labelled antiserum containing antibodies to a lysosomal enzyme (the antigen); the sites in the cell occupied by the latter appear as apple-green fluorescence when viewed in the fluorescence microscope. This method has also been modified for use at the level of the electron microscope by labelling the antibody with an electron opaque substance such as ferritin, a naturally occurring iron-rich protein.

In the absence of biochemical data, therefore, it is possible to identify lysosomes in tissues by a combination of histochemical and cytological methods. Frequently it is not possible to apply all these methods to the particular tissue under examination, but often sufficient biochemical data may also be available and permit conclusions as to the lysosomal nature of the particles under examination. Where biochemical study is possible it is desirable to combine this with histochemical observations. Only through such a combined approach has the study of lysosomes made such rapid progress.

## 2.1  Lysosomes in Protists, Animals and Plants

For ease of treatment I shall divide the remainder of this chapter dealing with the occurrence of lysosomes into three sections, corresponding to the three kingdoms of organisms, Protista, Animalia and Planta. The use of this classificatory system, proposed by Haeckel in 1866, is for convenience rather than an indication of total acceptance of this idea, but I should add that it is a scheme for which I have considerable sympathy. I shall take the protists to include those organisms of relatively simple biological organization such as algae, protozoa, fungi, slime-moulds and bacteria. Many readers may disagree with this scheme, but here we are interested not in semantics but rather in integrated biology.

### 2.1.1  Lysosomes in protists
*Protozoa.*  The feeding habits of the protozoa were well known by the latter part of the last century when the processes we now know as phagocytosis and pinocytosis had been frequently observed by Metchnikoff and others. Early observations on *Paramecium* had revealed the fusion of minute particles that accumulated neutral red dye with the food vacuoles, and a number of workers had suggested that in this manner digestive enzymes were conveyed to the food vacuole. Almost a century passed before these observations were extended and related to the lysosome concept. The use of methods of enzyme cytochemistry at the levels of the light and electron microscope have shown a number of lysosomal hydrolases in the food vacuoles of many protozoa including species of *Amoeba, Stentor, Tetrahymena, Ophryoglena, Paramecium* and other genera (Fig. 2.1). Most of the earlier contributions in this field, made by Müller's group working in Hungary, were reviewed by Müller *et al.* (1963). Combined morphological and cytochemical studies, particularly with the electron microscope, have

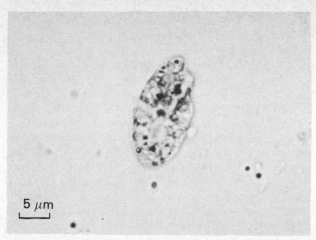

Fig. 2.1 Acid phosphatase activity in the food vacuoles of *Paramecium caudatum* as visualized in the light microscope. Fixed in neutral formalin and stained by the azo dye method of Grogg and Pearse (1952) — see Appendix.

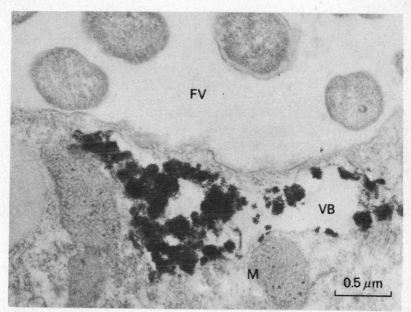

Fig. 2.2 Acid phosphatase localized in *Paramecium caudatum* by ultrastructural histochemistry. Thin section showing an empty vacuolar body (VB) with heavy lead phosphate deposits in proximity to a food vacuole (FV). M = mitochondrion. (From Esteve, J. C. (1970) *J. Protozool.*, 17, 23–34. Reprinted with permission of the Society of Protozoologists. Original print supplied by Dr J. C. Esteve.)

now revealed several types of membrane-bounded structures containing acid phosphatase which are thought to represent primary and secondary lysosomes, with the food vacuoles themselves being equivalent to a type of secondary lysosome. Biochemical characterization of some of these structures has been achieved in the giant amoeba, *Chaos chaos*, and in *Tetrahymena pyriformis*. Particularly detailed work by Müller *et al*. (1966) on the latter species demonstrated a distribution of five acid hydrolases in a type of particle with classic characteristics of lysosomes which was separable from mitochondria and peroxisomes by isopycnic centrifugation in a sucrose density gradient (Fig. 2.3). Current research has further demonstrated the heterogeneity of populations of lysosomes in protozoa by separating a low-density fraction rich in proteases, nucleases and acid phosphatase and low in activities of α-glucosidase, β-glucosidase and β-N-acetylglucuronidase from a high-density population containing the same enzymes but with reversed levels of activities. It is also known that the population of lysosomes in *Tetrahymena* undergoes profound alterations in its density distribution pattern during starvation or following interference with protein synthesis by the addition to cultures of chloramphenicol. *Tetrahymena pyriformis* also has the capacity to accumulate acridine dyes within the lysosome system.

Very recently there has been an upsurge of interest in the lysosomes and digestive processes of other protozoa, particularly the zooflagellates. Attention has centred around those parasitic flagellates of medical importance, notably *Trypanosoma, Leishmania* and *Trichomonas* and related genera. Several of the characteristic acid hydrolases occur in these organisms and cytochemical studies show that some of these are located in cytoplasmic granules. Although biochemical data are scarce the evidence favours the view that lysosomes exist in trypanosomatids and trichomonads, and that they play a role in intracellular digestion in the protozoa in general. We consider some further aspects of protozoan digestion in Chapter 3.

*Slime-moulds.* Recent biochemical, histochemical and ultrastructural observations on the slime-moulds have shown that acid hydrolases are associated with particles which satisfy some criteria for lysosomes. Particularly conspicuous in these organisms are food vacuoles rich in hydrolases and containing partially digested membranous materials and the remains of the bacteria on which they feed by heterophagy. There is also preliminary evidence suggesting that certain aspects of differentiation may be linked to autophagic phenomena in *Dictyostelium discoidium*. This is a field of very active research by Ashworth and Wiener (1973).

*Fungi.* Fungi secrete a wide range of extracellular enzymes which include acid hydrolases. Matile (1964) was able to show that *Neurospora crassa* secreted two constitutive proteases located in sub-cellular particles which

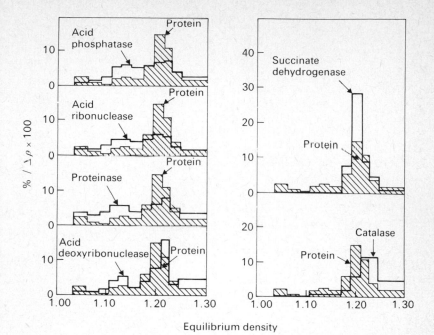

Equilibrium density

**Fig. 2.3** Distribution of proteins and marker enzymes for lysosomes, mitochondria and peroxisomes after subfractionation of *Tetrahymena* particles by means of density equilibration centrifugation in a sucrose gradient. Lysosomal enzymes equilibrate at the lower density ($\rho = 1.13$); mitochondrial enzymes at intermediate values ($\rho = 1.23$) and peroxisomal enzymes at dense parts of the gradient ($\rho = 1.23$). (From Müller, M. *et al.* (1966) *J. Cellular Physiol.*, **68**, 165–75. Courtesy of Wistar Institute Press.)

could be isolated by appropriate centrifugation techniques. Histochemical observations made in the author's laboratory demonstrated a particulate localization of acid phosphatase (Fig. 2.4), several esterases (Figs. 2.5 and 2.6) and β-galactosidase in the hyphae of *Botrytis cinerea* using light microscopy. Fungi were also shown to have the property of taking up acridine dyes (Fig. 2.7), 3,4-benzpyrene and neutral red which accumulated in lysosome-like particles. It was easily shown that the neutral red accumulated in structures which could be simultaneously stained for acid hydrolases, particularly acid phosphatase and indoxyl esterase, by procedures allowing direct microscopic observation of the staining progress. Work by other groups has now demonstrated the presence of acid phosphatase in vacuoles of fungi using ultrastructural histochemistry, Figs. 2.8 and 2.9. However, biochemical studies on the lysosomes of fungi have

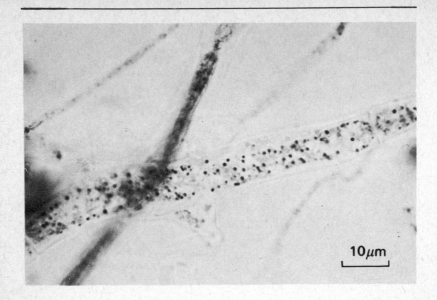

Fig. 2.4 Particulate localization of acid phosphatase by light microscopy in the hyphae of the fungus, *Botrytis cinerea*, revealed by the Gomori lead salt procedure after fixation in formol-calcium. (From Pitt, D. (1968) *J. gen. Microbiol.*, 52, 67–75. By permission of Cambridge University Press.)

not progressed rapidly owing to considerable difficulties experienced by physiologists in disrupting cells without simultaneously damaging the subcellular organelles. Attempts in the author's laboratory to overcome these problems by using enzymes which dissolve the hyphal walls and liberate the cell contents have not been very successful since the protoplasts so produced have little or no hydrolase activity compared to homogenates of the whole mycelium. Some success has been obtained by other workers in fractionating yeast cell organelles, particularly peroxisomes, and in addition Matile's group in Zürich has isolated and characterized vacuoles from yeast which were bounded by a single membrane, in this case the tonoplast, and contained a number of acid hydrolases that demonstrated latency. It is of historical interest that at least the secretory function of this organelle in yeasts and other fungi was suspected last century, and it is only recently that the affinities with the relatively newly discovered lysosome have been recognized. A further point of additional interest to this matter is the critical observation in *Pythium ultimum* by Grove *et al.* (1968) of a role of the Golgi apparatus in elaborating secretory vesicles which fuse with the plasma membrane. This provides support for the hypothesis of the Golgi-derived nature of lysosomes in general. If such

**Fig. 2.5** Lysosomal localization of acid phosphatase by light microscopy in the hyphae of the fungus *Botrytis cinerea*, stained by the standard coupling azo dye method of Grogg and Pearse (1952) — see Appendix. (From Pitt, D. (1968) *J. gen. Microbiol.*, **52**, 67—75. By permission of Cambridge University Press.)

observations could be extended by using cytochemical methods for hydrolases at the E.M. level our knowledge of the origin of lysosomes and their role in secretion would be greatly enlarged. It seems that fungi are particularly amenable to studies of this type, especially if procedures for sub-cellular fractionation could be improved.

*Algae.* Relatively little work has been carried out on the algae as a whole even though the group provides a range of excellent types of material for lysosome studies. In the last few years some interest has developed in research on the phytoflagellates. This is a group of simple plant-like flagellated microorganisms having diverse nutritional habits ranging through photoautotrophy, photoheterotrophy to heterotrophy. Additionally some species may be capable of feeding by taking in nourishment in soluble and particulate forms. There are classificatory problems with this group such that its members may be claimed or disclaimed by both protozoologists or phycologists. Since the present book is not primarily concerned with taxonomic matters I have chosen to include the phytoflagellates in the algae rather than the protozoa without feeling any need to justify the act.

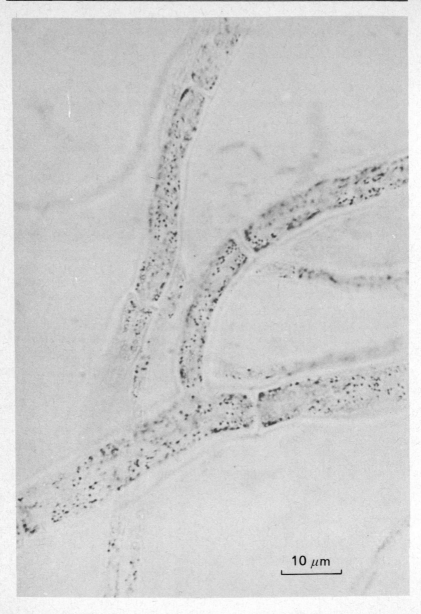

**Fig. 2.6** Localization by light microscopy of lysosomal thioacetic acid esterase in the hyphae of the fungus *Botrytis cinerea*. (From Pitt, D. (1968) *J. gen. Microbiol.*, 52, 67–75. By permission of Cambridge University Press.)

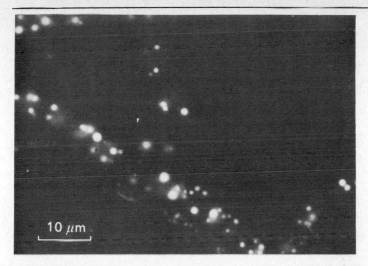

**Fig. 2.7**  Vital uptake of euchrysine 3R by lysosomes of the hyphae of the fungus *Botrytis cinerea*, as revealed by fluorescence microscopy. In the original, lysosomes (white) were orange against a black background. See Appendix for method.

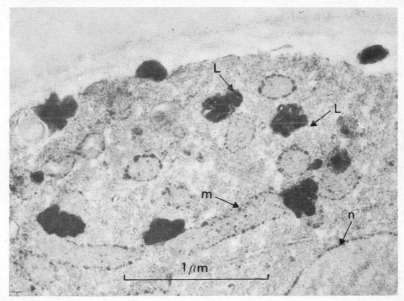

**Fig. 2.8**  Localization of acid phosphatase in lysosomes (L) in a thin section of a hypha of the fungus *Sclerotinia fructigena*, revealed by the acetoxymercuric aniline procedure of Smith and Fishman (1969) — see Appendix. Stain located in the nuclear membrane (n) and mitochondria (m) is not due to enzyme activity as it also appears in control sections. (Original print supplied by Dr E. C. Hislop.)

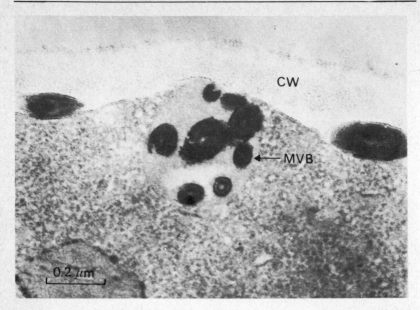

**Fig. 2.9** As in Fig. 2.8 but showing particulate localization of α-L-arabinofurano-sidase in a thin section of a hypha of *Sclerotinia fructigena*. It is believed that this represents a lysosomal location with reaction product also occurring in multivesicular bodies (MVB) and in the cell walls (CW). The possible involvement of this secreted enzyme in the rotting of infected apple fruits is discussed in Chapter 6. (From Hislop, E. C. *et al., J. gen. Microbiol*, 81, 79–99. By permission of Cambridge University Press. Original print supplied by Dr E. C. Hislop.)

There is some evidence that both extracellular and intracellular digestion occurs in this group, but only in *Euglena gracilis* have extensive observations been made on the lysosome system. Brandes's group has used biochemical, optical and electron microscope techniques on *Euglena*, in addition to a number of animal tissues, which have culminated in a continuing series of important publications on the role of lysosomes in cellular lytic processes. This work on *Euglena* has contributed to our general knowledge of the possible origins of primary lysosomes from flattened Golgi cisternae and the role of lysosomes in cellular autophagy which we shall consider later on. It is anticipated that expansion of the work on phytoflagellates will continue since this group of organisms provides useful tools for laboratory experimentation. In this context *Chlorogonium* has been used to elucidate the possible relationship between dictyosome vesicles and the origin of the vacuolar system, and *Chlamydomonas* is being used to investigate the role of lysosomes in aspects of phosphate regulation. Generally, however, the algae have been neglected except for observations on multivesicular bodies in *Chara*, and preliminary

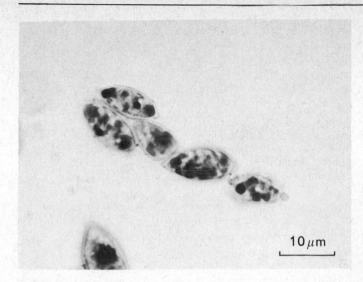

**Fig. 2.10** Acid phosphatase in the lysosomes of *Euglena gracilis*. Fixed in neutral formalin and stained by the azo dye method of Grogg and Pearse (1952) — see Appendix.

separations by zonal centrifugation of particulate acid hydrolases from the colourless alga *Polytomella caeca*

### 2.1.2 Lysosomes in Procaryotic Protists

In the foregoing treatment I have dealt with the evidence for the existence of lysosomes in eucaryotic protists. The situation is, of course, very different in the large group of procaryotic protists which includes the blue-green algae, bacteria and related forms. These organisms have no well-defined membrane system and are devoid of organelles comparable with those in eucaryotic organisms. It would seem, therefore, by definition that procaryotic types should be devoid of lysosomes even though they exhibit turnover of macromolecules and secrete extracellular enzymes. A speculative suggestion has been made by Matile that digestion in such organisms is located outside the cell membrane at the site where some hydrolases have been located. This has prompted the view that such cells are surrounded by an extracytoplasmic lysosome. There is little support for this theoretical extension of the lysosome concept at the present time although interesting contemporary views are expressed by Bissett (1973) on evolutionary aspects of the bacteria and the location of enzymes in and around the cell wall.

## 2.1.3 Lysosomes in Animals

Most of the fundamental work on lysosomes was carried out on liver and kidney, mainly of rats, such that these tissues have assumed the position of preferred experimental material in this field rather in the manner of *Escherichia coli* in molecular biology and *Drosophila* or *Neurospora* in genetics. As with all favourable experimental materials the more they are used the greater is the accumulation of knowledge and the more desirable becomes their use for elucidating further problems. Such a situation has its disadvantages and in the case of lysosomes it has meant that important information on the role of this organelle in other specialised tissue types has been slow in accumulating.

Lysosomes were discovered in rat liver and most of the early cytochemical observations subsequent to the discovery were carried out on this tissue. Immediate support for the biochemical observations came from studies on the 'droplet' fraction from rat kidney, a tissue which has also figured prominently in the development of cytochemical techniques for lysosomes. No other tissues have been examined as extensively as these organs of absorption and excretion since the original hypothesis envisaged lysosomes as primarily involved in the process of intracellular digestion. However, considerable attention has now been directed towards investigating secretory tissues, particularly the thyroid and prostate. Koenig has carried out extensive work on rat brain and on nerve tissues from a variety of mammalian species. Several investigators have concentrated on physiological and medical aspects of the functions of lysosomes in normal and diseased bone, cartilage and muscle. Considerable effort has been devoted to the epithelial cells of lung, intestine and male and female reproductive tracts, and to spermatozoa. A large volume of literature records the interest in lysosomes of the phagocytic cells of bone marrow, spleen and liver whilst particular attention has been directed towards the various blood cells, particularly leucocytes, and to the components of the reticulo-endothelial system, especially macrophages and lymphocytes. A wide variety of mammalian tissues has now been examined and it would seem that with the exception of red blood cells most cells possess a lysosomal system.

Apart from mammals a rapidly expanding literature exists which reports the occurrence and distribution of lysosomes in other vertebrate groups including birds, reptiles, amphibians and fish. The invertebrates have been examined in varying degrees of detail, mainly by light and electron microscopy, and the groups in which lysosomes have been found include arthropods, molluscs, annelids and echinoderms, some of the minor coelomate phyla, platyhelminths and nematodes, and the lower metazoa, particularly the sponges. It would seem that almost every animal tissue that has been investigated has yielded some evidence for the existence of lysosomes, but apart from mammalian species and some other tissues of special interest, e.g. chick fibroblasts, amphibia, insects and sponges, the

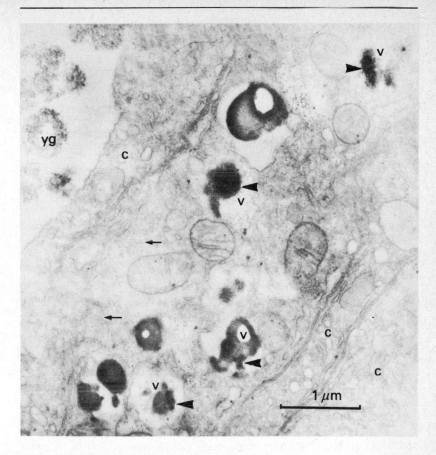

Fig. 2.11    Thin section of the digestive cell of the slug, *Agriolimax reticulatus*, stained for acid phosphatase activity by the post-coupling azo dye procedure. Reaction product in the form of fine dense particles can be seen concentrated within vacuoles (v). The particles (large arrows) appear superimposed upon the dense homogeneous contents of the vacuoles. Small arrows show the association of reaction product with the cytoplasm which is rich in free ribosomes. The cytoplasm of adjacent cells (c) and the so-called yellow granules (yg) are negative. (From Bowen, I. D. (1971) *J. Microscp.*, 94, 25–38. By courtesy of the *Journal of Microscopy*. Original print supplied by Dr I. D. Bowen.)

biochemical evidence is sparse. It seems likely that any desire to fulfil the biochemical criteria could be satisfied.

The brevity of this section does not indicate a corresponding lack of scientific interest or a paucity of information on the distribution and occurrence of lysosomes in animals, but, on the contrary reflects the breadth and depth of the available facts, a brief review of which would fill this book several times over. Interested readers wishing to obtain information relating to the occurrence of lysosomes in particular tissues or species are referred either to the general reviews of Straus (1967) and Gahan (1967), or to the selected specialist references given at the end of this chapter.

### 2.1.4 Lysosomes in Plants

Uncertainty exists regarding the occurrence in plant cells of organelles equivalent to animal lysosomes. This situation seems to have arisen through strict adherence to the original biochemical definition which, as we have already seen, has become modified in the light of expanding knowledge of the heterogeneity of lysosomes. It will emerge from the following account, that striking similarities exist between plant and animal lysosomes which suggest that it is perhaps logical to encompass all related structures within the lysosome concept.

Early studies by plant biochemists and physiologists on the isolation of lysosomes from plants met with failure since attempts sought to apply directly those methods developed in animal work without due regard to the special properties of plant cells. Sub-cellular fractionation of plant tissues is fraught with technical difficulties related to the high osmolarity of most plant tissues, high organic acid concentrations in the cells, the reaction of polyphenol oxidases with substrates following the disorganization of compartmentation during homogenization to produce free radical generating systems which destroy cell membranes. Recent work by Galliard (1971) on potato tubers implicates endogenous lipoxygenase and phospholipase in membrane and organelle destruction during tissue homogenization and fractionation. In addition the methods necessary to disrupt cell walls require shearing forces of magnitudes which also disrupt most of the sub-cellular components. The first successful isolation of a particulate fraction rich in hydrolases was achieved by Harrington and Altschul in 1963 who obtained a light mitochondrial fraction from onion seedlings, the sedimentable enzymes of which were released to the supernatant fluid fraction after treatment with the membrane-destroying detergent Triton X-100. The successes of the plant cytochemists at this time were in marked contrast to the discouraging results from cell fractionation studies and early observations reported the cytochemical localization of acid phosphatase by light microscopy in root tips of a number of species, though such activity was initially attributed to the mitochondria. Very successful cytochemical studies were initiated in Poland in the early 1960s in which Walek-Czernecka and her co-workers revealed several particulate lysosomal hydrolases including phosphatase, sulphatase, $\beta$-glucuronidase, $\beta$-galactosidase and deoxyribonuclease in onion bulb-scale epidermis, root tips of grasses and of larch. Lysosomes of the tissues from a number of plant species are shown in Figs. 2.12 to 2.15. Cytochemical work has been

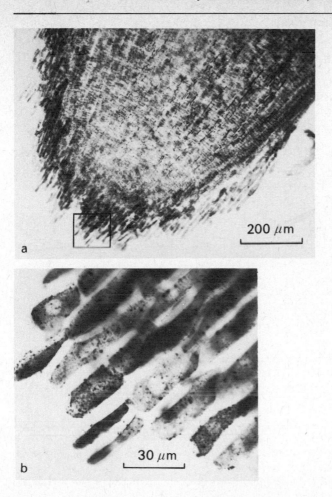

**Fig. 2.12** (*a*) Distribution of acid phosphatase activity in pea root tips (*Pisum sativum*) as shown by light microscopy. Fixed in neutral formalin and stained by the simultaneous-coupling azo dye method of Grogg and Pearse (1952) — see Appendix. (*b*) Higher magnification of the area inset in (*a*) showing particulate localization of the enzyme in cells of the root cap.

continued by several workers in Poland who revealed similar lysosome locations in anthers, pollen grains and pollen tubes of many species. Important contributions were also made by Gahan, working in London, who was able to show particulate hydrolase locations with the light microscope by using the newer azo dye methods, and also demonstrated structure-linked latency in *Vicia faba* root tips by cytochemical methods.

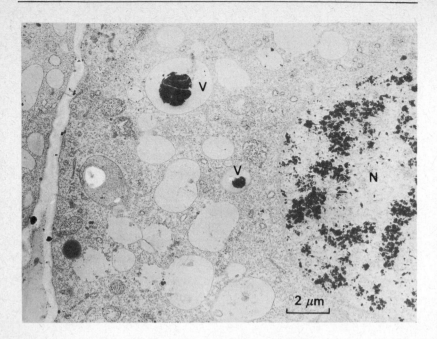

**Fig. 2.13** Electron micrograph of a thin section of a cell from the root tip of pea (*Pisum sativum*). Fixed in glutaraldehyde and stained by the Gomori lead salt procedure for acid phosphatase. Reaction product is deposited within vacuoles (V) and in the nucleus (N). (From Sexton, R. *et al*. (1971) *Protoplasma*, 73, 417–48. © Springer-Verlag, 1971. Original print supplied by Dr J. L. Hall.)

During this period some initial progress was made towards the cytochemical localization of acid phosphatase by the Gomori lead salt method using ultrastructural techniques. However, the high content of polyphosphates in plant tissues makes the method notoriously unreliable. Recent use of rigorously controlled Gomori procedures have now confirmed the ultrastructural localization of acid phosphatase in membrane-bounded vesicles which have characteristics of lysosomes (Figs. 2.13 and 2.14). Attempts to use the much more reliable and specific azo dye procedures have encountered little success in our hands using leaf, stem and root tissues since the reaction products, which are clearly seen at particulate sites in the light microscope, are soluble in those dehydrating fluids required to give good tissue preservation for electron microscopy. This is a field of endeavour in which considerable progress is urgently required. Contemporary ideas on the occurrence and histochemistry of plant lysosomes are reviewed by Gahan (1973).

These achievements by the histochemists stimulated further interest in

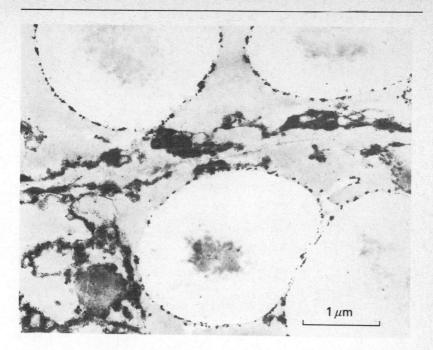

**Fig. 2.14** Electron micrograph of a thin section of cortical cells of the root of maize (*Zea mays*) stained for acid phosphatase by the Gomori procedure. Reaction product is localized at the periphery of the vacuoles (lysosomes). (From Hall, J. L, and Davie, C. A. M. (1971) *Ann. Bot.*, **35**, 849–55. By permission of the *Annals of Botany*. Original print supplied by Dr J. L. Hall.)

plant lysosomes, and striking early successes were obtained by Matile's group in Zürich which developed techniques permitting the isolation of particles from tobacco and maize seedlings that contained several acid hydrolases characteristic of animal lysosomes, e.g. protease, acid phosphatase, carboxylic esterase and ribonuclease. The discovery of an amylase in these structures illustrated a difference from animal lysosomes and a similarity to the situation in some protozoa. Important lysosomal enzymes of animal tissues such as β-glucuronidase and phospholipase were not found in these plant organelles. Present work by the author has shown sedimentable acid hydrolases in leaves, shoots and tissue culture cells of a number of species. Current findings with lysosome fractions from potato shoots (Fig. 2.16) showed peaks of acid hydrolase activity containing acid phosphatase, phosphodiesterase, ribonuclease, carboxylic esterase and β-glycerophosphatase which were well separated from peaks of mitochondrial and glyoxysomal marker enzymes (Fig. 2.17). A heavy lysosome fraction was isolated with particle diameters of 0.1–1.6 μm and a density

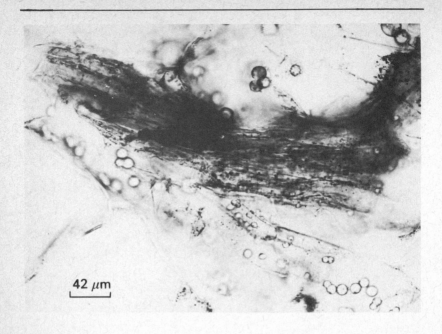

42 μm

**Fig. 2.15** Section of potato tuber tissue stained for acid phosphatase by the standard coupling azo dye method after fixation in neutral formalin. Note many small particles (lysosomes) especially in the vascular tissues with some diffuse cytoplasmic staining. (From Pitt, D. and Coombes, C. (1968) *J. gen. Microbiol.*, 53, 197–204. By permission of Cambridge University Press.)

of $1.10 \text{ g cm}^{-3}$ which contained autophagic vacuoles (Fig. 2.18). This fraction was distinguishable from a low density fraction ($1.07 \text{ g cm}^{-3}$) of smaller particles ($0.025–0.6 \mu m$ in diameter) with higher levels of hydrolase activity (Fig. 2.19). In this work the interesting observation was made, and has been referred to in Chapter 1, p. 18, that the acid phosphatase activity appeared to be structurally bound to the membrane through a phospholipid bond, and that the two populations of lysosomes were characterized by different molecular forms of acid phosphatase.

So far we have referred only to observations made on higher plants. Recently, however, Coulomb (1969) has reported one of the few attempts to separate lysosomes from lower plants when he demonstrated sedimentable hydrolases from the fronds of the fern *Asplenium fontanum*.

As mentioned earlier (Chapter 1, p. 16) the origins and interrelationships of the various hydrolase-containing particles in plants have not been extensively examined, but initial observations by electron microscopy have shown broad similarities to the proposed situation in animals and imply a dictyosomal origin of provacuoles, although some workers favour an

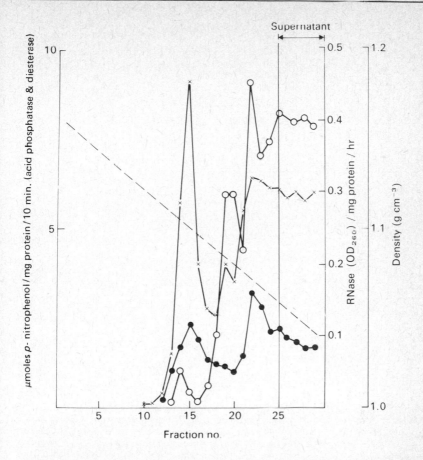

**Fig. 2.16** Distribution of acid hydrolases from potato shoots in a linear Ficoll gradient after ultracentrifugation. x —————— x acid phosphatase; ● —————— ● phosphodiesterase; ○ —————— ○ ribonuclease; —————— density. Activity is expressed as units/mg protein. Note two main peaks of hydrolase activity occurring at fractions of densities 1.10 g cm⁻³ and 1.07 g cm⁻³ which correspond to the 'heavy' and 'light' lysosome fractions respectively. (From Pitt, D. and Galpin, M. (1973) *Planta (Berl.)*, 109, 233–58. © Springer-Verlag, 1973.)

ER-derived origin for plant lysosomes. Current views on the origin of the vacuoles in plants are discussed by Berjak (1972) who believes that the vacuole system represents a major component of the lysosomal system of plant cells. Considering the evidence as a whole there now seems little doubt from the combined biochemical, cytochemical and ultrastructural studies that the acid hydrolase-containing particles in plants have strong

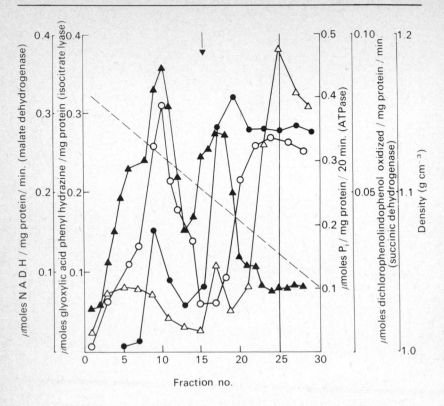

Fig. 2.17   Fractionation of potato shoot organelles by ultracentrifugation. Comparative distributions of mitochondrial, glyoxysomal and lysosomal marker enzymes in a linear Ficoll gradient. ● ————— ● ATPase; ○ ————— ○ malate dehydrogenase; ▲ ————— ▲ succinic dehydrogenase; △ ————— △ isocitrate lyase; ————— density; position of acid phosphatase marker (i.e. 'heavy' lysosomes). Enzyme activity of fractions is expressed as units/mg protein. (From Pitt, D. and Galpin, M. (1973) *Planta (Berl.)*, **109**, 233–58. © Springer-Verlag, 1973.)

affinities with the lysosomes of animals and protists. Indeed we shall see later that the vacuolar system forms part of the intracellular digestive system of plant cells certainly through autophagy, and now that pinocytosis is a demonstrable feature of plant cells, perhaps through heterophagy also.

Before closing this section I need to refer briefly to a further group of plant cell organelles, namely the spherosomes, which have been known for many years and were studied by early plant cytologists. Frey-Wyssling observed the ontogeny of spherosomes and concluded that they were budded from the ER. The studies of Dangeard (1919) and Frey-Wyssling

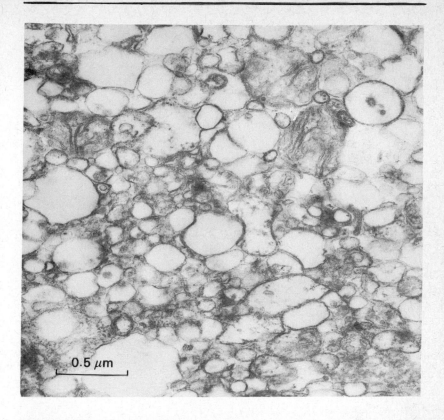

**Fig. 2.18** Electron micrograph of a thin section of the 'heavy' lysosomal fraction isolated from potato shoots. (From Pitt, D. and Galpin, M. (1973) *Planta (Berl.)*, 109, 233—58. © Springer-Verlag, 1973.)

*et al.* (1963) showed that the young spherosomes were gradually transformed into lipid droplets on aging, but it was not clear at that time if lipid material was actually synthesized at these sites or if it merely accumulated there after synthesis elsewhere. Our understanding of the nature of spherosomes is confused by the observations on impure subcellular fractions since it is apparent that some of the earlier preparations of lysosomes obtained by differential centrifugation must have been heavily contaminated with spherosomes because not only did they contain acid hydrolases but also incorporated $^3$H-acetate into lipid. For a while it became accepted that lysosomes were indeed the spherosomes of ancient knowledge. Observations in our laboratory reveal that the lysosome preparations from leaves and shoots were not significantly contaminated

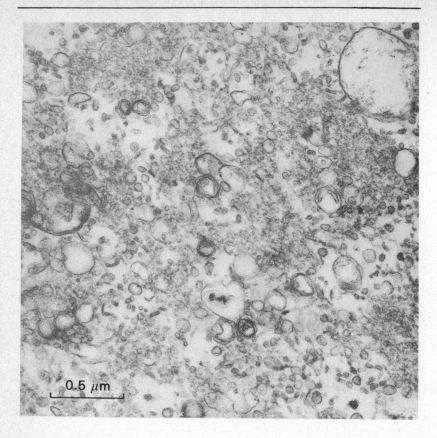

**Fig. 2.19** Electron micrograph of a thin section of the 'light' lysosomal fraction isolated from potato shoots.

by spherosomes since the latter accumulated on the surface of density gradients and were well separated from the other major organelles. It is possible that some of the acid hydrolase activity found associated with these structures could be an artefact due to adsorption of enzyme from the soluble fraction of cell homogenates. Also there are no reports of auto-phagic bodies associated with purified spherosomes nor do such prepara-tions show the characteristic heterogeneity of lysosome fractions. Further-more, extensive work by Yatsu and Jacks over recent years has thrown new light on the structure, function and fate of spherosomes and has indi-cated that they are a group of morphologically distinct organelles. The most recent work has shown that the 'single-line' membrane bounding the spherosomes is in fact a half unit-membrane, and strengthens the view that

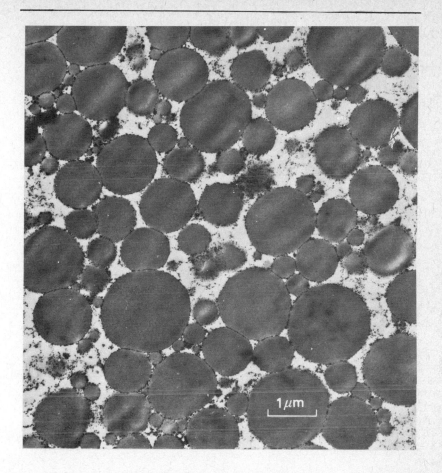

Fig. 2.20  Electron micrograph of a thin section of isolated spherosomes from pea-nuts. Major differences in respect of morphology, nature of the membrane and its lipid content are seen when comparing these organelles with the lysosomes in Figs. 2.13 and 2.14. (From Yatsu, L. Y. and Jacks, T. J. (1972) *Plant Physiol.*, 49, 937—43. Courtesy of the American Society of Plant Physiologists. Original print supplied by Dr L. Y. Yatsu.)

plant spherosomes and the lipid droplets of animals, e.g. adipose droplets, have strong affinities. It now seems appropriate on the available evidence to consider plant spherosomes as structures which are quite distinct from lysosomes and this being so we shall not consider the former again in the present text. The fine structure of isolated spherosomes is shown in Figs. 2.20 and 2.21 and is in marked contrast to that recognized for lysosomes.

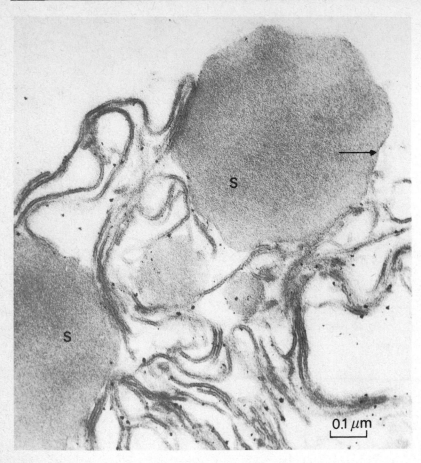

Fig. 2.21   Ultrastructure of spherosomes from peanut (*Arachis hypogea*). A high resolution electron micrograph of a spherosome preparation showing the single-line membrane (at arrow) which is barely distinguishable due to the osmiophilia of the spherosome content (S). The nature of this type of membrane is in marked contrast to the membrane typical of lysosomes. (From Yatsu, L. Y. and Jacks, T. J. (1972) *Plant Physiol.*, 49, 937–43. Courtesy of the American Society of Plant Physiologists. Original print supplied by Dr L. Y. Yatsu.)

## Suggestions for further reading

Protozoa
EECKHOUT, Y. (1973) Digestion and lysosomes in zooflagellates, in *Frontiers of Biology*, 29. *Lysosomes in Biology and Pathology*, vol. 3, ch. 1, pp. 3–17, ed. J. T. Dingle. North-Holland Publishing Co., Amsterdam and London.

MÜLLER, M. (1971) Lysosomes in *Tetrahymena pyriformis* II. Intracellular distribution of several acid hydrolases, *Acta Biol. Acad. Sci. Hung.*, 22, 179–86.
MÜLLER, M., BAUDHUIN, P. and de DUVE, C. (1966) Lysosomes in *Tetrahymena pyriformis* I. Some properties and lysosomal localization of acid hydrolases, *J. Cell. Comp. Physiol.*, 68, 165–76.
MÜLLER, M., RÖHLICH, P., TÓTH, J. and TÖRÖ, I. (1963) Fine structure and enzymic activity of protozoan food vacuoles, in *Lysosomes*. C.I.B.A. Foundation Symposium on Lysosomes, pp. 201–16, eds. A. V. S. de Reuck and M. P. Cameron. J. and A. Churchill Ltd., London.

**Slime-moulds**
ASHWORTH, J. M. and WEINER, E. (1973) The lysosomes of the cellular slime mould *Dictyostelium discoideum*, in *Frontiers of Biology*, 29. *Lysosomes in Biology and Pathology*, vol. 3, ch. 3, pp. 38–48, ed. J. T. Dingle. North-Holland Publishing Co., Amsterdam and London.
KAZAMA, F. W. and ALDRICH, H. C. (1972) Digestion and the distribution of acid phosphatase in the myxamoeba *Physarum flavicomum*, *Mycologia*, 64, 529–38.

**Fungi**
GROVE, S. N., BRACKER, C. E. and MORRÉ, D. J. (1968) Cytomembrane differentiation in the endoplasmic reticulum—Golgi apparatus—vesicle complex, *Science*, 161, 171–3.
MATILE, P. (1964) Die Funktion proteolytischer bei der Proteinaufnahme durch *Neurospora crassa*, *Naturwissenschaften*, 51, 489–90.
MATILE, P. (1969) Plant lysosomes, in *Frontiers of Biology*, 14A. *Lysosomes in Biology and Pathology*, vol. 1, ch. 15, pp. 406–30, eds. J. T. Dingle and H. B. Fell. North-Holland Publishing Co., Amsterdam and London.
MATILE, P. and WIEMKEN, A. (1967) The vacuole as the lysosome of the yeast cell, *Archiv. für Mikrobiol.*, 56, 148–57.
PITT, D. (1968) Histochemical demonstration of certain hydrolytic enzymes within cytoplasmic particles of *Botrytis cinerea* Fr., *J. gen. Microbiol.*, 52, 67–75.
WILSON, C. L., STIERS, C. L. and SMITH, G. S. (1970) Fungal lysosomes or spherosomes, *Phytopathology*, 61, 216–27.

**Algae**
AARONSON, S. (1973) Digestion in phytoflagellates, in *Frontiers of Biology*, 29. *Lysosomes in Biology and Pathology*, vol. 3, ch. 2, pp. 18–37, ed. J. T. Dingle. North-Holland Publishing Co., Amsterdam and London.
BRANDES, D. (1965) Observation on the apparent mode of formation of 'pure' lysosomes, *J. Ultrastruct. Res.*, 12, 63–80.
BRANDES, D., BEUTOW, D. E., BERTINI, F. J. and MALKOFF, D. B. (1965) Role of lysosomes in cellular lytic processes. I. Effect of carbon starvation in *Euglena gracilis*, *Exptl. Mol. Pathol.*, 3, 583–92.
UEDA, K. (1966) Fine structure of *Chlorogonium elongatum* with special reference to vacuole development, *Cytologia*, 31, 161–72.

**Animals — General References**
GAHAN, P. B. (1967) Histochemistry of lysosomes, *Int. Rev. Cytol.*, 21, 1–63.
STRAUS, W. (1967) Lysosomes, phagosomes and related particles, in *Enzyme Cytology*, ch. 5, pp. 239–319, ed. D. B. Roodyn. Academic Press, London and New York.

**Mammals**
de DUVE, C., PRESSMAN, B. C., GIANETTO, R., WATTIAUX, R. and APPELMANS, F. (1955) Tissue fractionation studies. 6. Intracellular distribution patterns of enzymes in rat-liver tissue, *Biochem. J.*, 60, 604–17.

KOENIG, H. (1969) Lysosomes in the nervous system, in *Frontiers of Biology*, 14B. *Lysosomes in Biology and Pathology*, vol. 2, ch. 6, pp. 109–62, eds. J. T. Dingle and H. B. Fell. North-Holland Publishing Co., Amsterdam and London.
WATTIAUX, R. (1969) Biochemistry of lysosomes, in *Handbook of Molecular Cytology*, pp. 1159–78, ed. A. Lima-de-Faria. North-Holland Publishing Co., Amsterdam and London.

Other Vertebrates
HOPKINS, C. R. (1969) The fine structural localization of acid phosphatase in the prolactin cell of the teleost pituitary following stimulation and inhibition of secretory activity, *Tissue and Cell*, 1, 653–71.
WEBER, R. (1969) Tissue involution and lysosomal enzymes during anuran metamorphosis, in *Frontiers of Biology*, 14B. *Lysosomes in Biology and Pathology*, vol. 2, ch. 15, pp. 437–61, eds. J. T. Dingle and H. B. Fell. North-Holland Publishing Co., Amsterdam and London.

Invertebrates
BOWEN, I. D. and DAVIES, P. (1971) The fine structural distribution of acid phosphatase in the digestive gland of *Arion hortensis* (Fr.), *Protoplasma*, 73, 73–81.
LOCKSHIN, R. A. (1969) Lysosomes in insects, in *Frontiers of Biology*, 14A. *Lysosomes in Biology and Pathology*, vol. 1, ch. 13, pp. 364–91, eds. J. T. Dingle and H. B. Fell. North Holland Publishing Co., Amsterdam and London.
SUMNER, A. T. (1969) The distribution of some hydrolytic enzymes in the cells of the digestive gland of certain lamellibranchs and Gastropods, *J. Zool.* (Lond.), 158, 277–91.

Lower Metazoa
TESSENOW, W. (1969) Lytic processes in development of freshwater sponges, in *Frontiers of Biology*, 14A. *Lysosomes in Biology and Pathology*, vol. 1, ch. 14, pp. 392–405, eds. J. T. Dingle and H. B. Fell. North-Holland Publishing Co., Amsterdam and London.
TIFFON, Y., RASMONT, R., DEVOS, L. and BOUILLON, J. (1973) Digestion in lower metazoa, in *Frontiers of Biology*, 29. *Lysosomes in Biology and Pathology*, vol. 3, ch. 4, pp. 49–68, ed. J. T. Dingle. North-Holland Publishing Co., Amsterdam and London.

Plants
BERJAK, P. (1972) Lysosomal compartmentation: Ultrastructural aspects of the origin, development and function of vacuoles in root cells of *Lepidium sativum*, *Ann. Bot.*, 36, 73–81.
HARRINGTON, J. F. and ALTSCHUL, A. M. (1963) Lysosome-like behaviour in germinating onion seeds, *Fedn. Proc. Fedn. Am. Socs. exp. Biol.*, 22, 475.
DANGEARD, P. A. (1919) Sur la distinction du chondriome des auteurs vacuome, plastidome et sphérome, *C.R. hebd. Seanc. Acad. Sci.*, Paris, 169, 1005–10.
FREY-WYSSLING, A., GRIESHABER, E. and MÜHLETHALER, K. (1963) Origin of spherosomes in plant cells, *J. Ultrastruct. Res.*, 8, 506–16.
GAHAN, P. B. (1967) Lysosomes, in *Plant Cell Organelles*, ch. 13, pp. 228–38, ed. J. B. Pridham. Academic Press, London and New York.
GAHAN, P. B. (1973) Plant lysosomes, in *Frontiers of Biology*, 29. *Lysosomes in Biology and Pathology*, vol. 3, ch. 5, pp. 69–85, ed. J. T. Dingle. North-Holland Publishing Co., Amsterdam and London.
MATILE, P. (1969) Plant lysosomes, in *Frontiers of Biology*, 14A. *Lysosomes in Biology and Pathology*, vol. 1, ch. 15, pp. 406–30, eds. J. T. Dingle and H. B. Fell. North-Holland Publishing Co., Amsterdam and London.

MATILE, P. and MOOR, H. (1968) Vacuolation: Origin and development of the lysosomal apparatus in root-tip cells, *Planta (Berl.)*, 80, 159—75.

PITT, D. and GALPIN, M. (1973) Isolation and properties of lysosomes from dark-grown potato shoots, *Planta (Berl.)*, 109, 233—58.

YATSU, L. Y. and JACKS, T. J. (1972) Spherosome membranes. Half unit-membranes, *Plant Physiol.*, 49, 937—43.

General

BISSET, K. (1973) This prokaryotic—eukaryotic business, *New Scientist*, 57, 296—8.

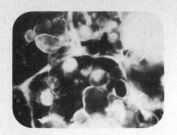

Chapter 3

# Physiological Functions of Lysosomes: Role of Lysosomes in Intracellular Digestion

Although lysosomes are now known to be involved in several sub-cellular phenomena, their activities are all basically related to the process of intracellular digestion.

The ability to form cytoplasmic vacuoles is a characteristic of eucaryotic cells which is completely absent in cells of procaryotic organisms. Most cells have the capacity to take in individual molecules, but in those eucaryotic cells devoid of cell walls, i.e. protozoa and animals, there exists the extensive facility for carrying out the process of *endocytosis* whereby bulk intake of exogenous materials may be achieved. Endocytosis is a generalized term embracing two phenomena; the incorporation of droplets of liquid into the cell from the outside by the process of *pinocytosis*, and the ingestion of solid matter by *phagocytosis*. Both are energy-requiring processes probably involving initial adsorption of material onto the cell surface with subsequent extension of the surface membrane around the particle, in an amoeboid fashion, to incorporate the droplet or the particle into the cytoplasmic vacuole or *phagosome* wherein digestion may occur through the agency of lysosomal enzymes which are discharged into the vacuole. These processes characterize the mode of nutrition amongst those protists such as protozoa, slime-moulds and some flagellates which form *food vacuoles*. This form of nutrition also prevails in some protists with cell walls, e.g. *Euglena*, when the absorption process takes place from the base of a gullet. The products of digestion then pass out of the vacuole by a little understood process to the cytoplasm, where they are utilized in metabolic processes. This method of *intracellular digestion* is also found in some invertebrate animals that do not have a gastrointestinal tract. However, in more highly developed animals *extracellular digestion* of solid material occurs in the gastrointestinal tract following the action of secreted digestive enzymes, with the soluble products being absorbed through the gut wall. However, the process of intracellular digestion is not a characteristic exclusive to lower organisms. It is now recognized as a feature of most eucaryotic cells which permits the digestion, by enzymes produced in the same cells, of molecules and pieces of cytoplasmic material of endogenous origin in the process of *autophagy*. By means of this process the re-cycling of components within the cell is achieved in the phenomenon known as *turnover*. Most animal cells show, in addition, the capacity to carry out some degree of pinocytosis and phagocytosis during which extracellular foreign matter is taken into the cell in *heterophagy* and may be subsequently degraded via the intracellular digestive system. In those animals which carry out extracellular digestion in cavities such as the stomach and intestine the role of phagocytosis is no longer primarily nutritional but assumes a *protective* role by providing a mechanism for engulfing and destroying foreign materials, particularly invading microorganisms. Such a role was established in 1882 by Eli Metchnikoff's penetrating observations on starfish larvae. He noted that within the transparent body of the larva mobile amoeboid cells circulated which surrounded a foreign object (in this case a rose thorn) that he introduced into the tissues. This he interpreted as a reaction of the cells of the starfish larva to foreign matter which we now recognize as the basis of inflammation. Amidst considerable controversy other work quickly confirmed that *phagocytosis* ('devouring by cells'), as he termed the process, was a general mechanism of defence against invading microorganisms in nearly all animal cells. It is now realized that both phagocytes and *antibodies* play important roles in defence mechanisms and that there are interrelationships between the two systems to the extent that one function of antibodies is to increase the efficiency of phagocytosis. Until recently heterophagy was not considered a marked attribute of plant cells but contemporary studies by Mahlberg (1972) confirmed the much earlier belief that these cells *can* perform pinocytosis, but the physiological significance of the process has not been evaluated at present. The presence of the cell wall would seem to preclude phagocytosis by plant cells of all except the smallest macromolecules but the process has been demonstrated in isolated plant protoplasts using much larger particles, e.g. latex. As with pinocytosis the physiological significance of phagocytosis is not yet apparent, but it may be important in the uptake of some plant viruses.

Once the surface layers of the animal body have been penetrated, the circulatory system could provide a convenient means of spreading the infective agents. However, within the circulatory system there are important defence mechanisms composed of, apart from antimicrobial substances, phagocytic cells, namely two types of white blood cells, *neutrophils* and *monocytes* which have the capacity to leave the bloodstream and phagocytose invading microorganisms and dead tissues. Another type of phagocyte, known as the *fixed macrophage*, lines the lumens of certain organs of the circulatory system including the spleen, liver, bone marrow and the lymph system. An additional type of phagocyte, the *wandering macrophage* or *histiocyte*, occurs throughout the connective tissues. The fixed macrophages not only line, i.e. they are endothelial, but also form a reticular network throughout certain organs of the circulatory system and are known collectively as the reticulo-endothelial (RE) system. All the circulatory fluid of the animal body must pass through the RE system and is thereby cleansed of most of the invading microorganisms and worn-out cells such as erythrocytes. Those that survive the purgative effect of this system may be engulfed by the neutrophils or monocytes of the blood. Thus, the combined action of the antimicrobial substances and phagocytes form the front line of the animal body defences. The situation in plants is very different in that the cell wall plays a key role in excluding pathogens. Infections that are established must overcome the antimicrobial effect of *phytoalexins* produced at the site of infection. A third line of defence in plants is through the existence of the genetically-controlled *hypersensitive* response whereby infected tissues, and those in the immediate vicinity of infection, die, and obligate pathogens, which depend on the availability of living tissues for their nutrients, are starved and eventually die also with the net effect being small necrotic spots on the tissues as evidence of such activities.

## 3.1 Intracellular Digestion

I have already outlined events involved in the formation of the various types of lysosome (Fig. 1.7), and now I wish to examine in more detail some pieces of work which provided evidence supporting the notion that lysosomes were the seat of intracellular digestion not merely in protozoa, but also in the metazoa and plants. The literature concerning lysosomes and related particles is huge and varying terminologies are used which have created a state of confusion. The term 'lysosome' was introduced originally to describe those cytoplasmic granules containing several acid hydrolases which showed latency. Subsequent work revealed great heterogeneity amongst the lysosomes and it is now apparent that such polymorphism is related to their various physiological functions within cells. Thus, several specific types of lysosome may be distinguished, viz. digestive vacuoles, storage granules, autophagic and residual bodies, all of which contain lysosomal enzymes at some stage of their existence although the latter ultimately show no active hydrolase content. This heterogeneity lead various workers to use special distinguishing terminology to describe the appearance and function of the particles in the particular system under their scrutiny. The term *phagosome* was introduced to describe vacuoles arising through endocytosis and which at some later stage incorporated

Table 3.1   Nomenclature and interrelationships of components of the lysosome
system

| Term | Synonym | Characteristics |
|------|---------|-----------------|
| *Primary lysosome* | Cytosome; pure lysosome; proto-lysosome; young lysosome | Organelles derived from Golgi and/or ER. Single membrane, contain acid hydrolases |
| *Secondary lysosome* | Phago-lysosome; digestive vacuole; old-lysosome; food vacuole (of protozoa) | Body produced on fusion of a primary lysosome with a vacuole containing material to be digested |
| *Autophagic vacuoles* | Cytolysome; segresome; cytosegresome; composite body; secondary lysosome | Single or double membrane. Contain recognizable cytoplasmic components in various stages of digestion |
| *Multivesicular body* | Late phago-lysosome; secondary lysosome | Usually bounded by a single membrane. Contents vacuolated and may represent a late secondary lysosome |
| *Residual body* | Dense-body; post-lysosome; storage-body; telolysosome | Usually single membrane bounded, may be double. Contains indigestible remains. Often whorled and containing myelin, etc. Low or no hydrolase activity |

lysosomal enzymes, after which they were recognized as *phago-lysosomes*. Enlarged lysosomes were also found which contained various cytoplasmic components, e.g. ER, mitochondria, membraneous residues, and were variously named as *'cytolysomes', 'autophagic bodies'* or *'cytosegresomes'*. Several attempts have been made to standardize the terminology, notably by Gordon, Miller and Bensch (1965), but no acceptable unifying scheme has yet emerged. In order that students might comprehend the diverse terminology in the literature Table 3.1 lists the various synonyms; where possible we shall use those terms in the left-hand column.

## 3.2   Heterophagic Functions of Lysosomes

I have already mentioned that the phagocytic nature of the mode of nutrition of protozoa was well established by the end of the last century long before the realization that lysosomes were involved in digestion. Classic work on amoebae by Holter (1959; 1961) and of Chapman-

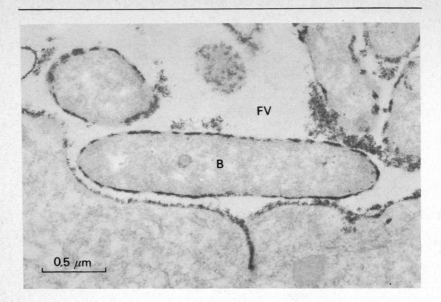

Fig. 3.1 A food vacuole of *Paramecium caudatum* containing deposits of lead phosphate, indicative of acid phosphatase activity, located at the periphery of the engulfed bacterium (B) and of the vacuole (FV). (From Esteve, J. C. (1970) *J. Protozool.*, 17, 23—34. Reprinted with permission of the Society of Protozoologists. Original print supplied by Dr J. C. Esteve.)

Andresen (1962) at the Carlsberg Laboratories heralded the onset of the current phase of knowledge of the nature of endocytic uptake by cells in general. Recent insight into the intracellular features of digestive processes in protozoa has been provided by Esteve (1970) who followed the fate of ingested materials, including bacteria, in *Paramecium caudatum* using ultrastructural histochemistry. These studies confirmed and extended the existing views on the fate of ingested materials made in a wide range of protozoan species. In this species acid phosphatase was located in the ER, Golgi apparatus, food vacuoles (Fig. 3.1), autophagic bodies and vacuolar and dense bodies. It seems that the enzyme originated from both the ER and Golgi apparatus, and reached the food vacuoles from either source. Digestion of engulfed materials, bacteria of exogenous origin and enclosed organelles arising through autophagy, resulted in the production of dense bodies, rich in acid hydrolases, which may have contained end-products of digestion due for assimilation by the cell. Conversely, the dense bodies may have been destined for subsequent fusion with the vacuole. It is diffi-cult to be certain on this matter and there are dangers of drawing the wrong conclusions from static E.M. observations, but the available evidence seems to favour the opinion that dense bodies *arise* from the food

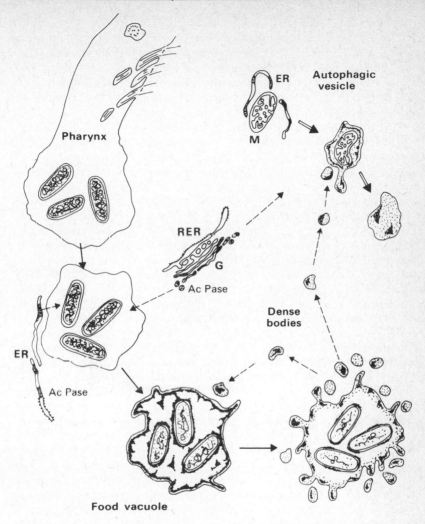

**Fig. 3.2** Schematic representation of the possible relationships between the acid phosphatase-positive structures in *Paramecium caudatum*. The enzymic activity of the food vacuole is probably derived from the ER and small Golgi vesicles. In later stages of digestion dense granules may arise from the food vacuoles and act as storage sites for the enzyme(s), or they may fuse with young food vacuoles and autophagic vacuoles (i.e. they may act in a cyclic fashion in digestive bouts). The autophagic vacuoles are probably formed by the incorporation of cytoplasmic organelles into a structure similar to an ergastoplasmic cisterna. It is believed that the autophagic vacuoles may ultimately be converted to an additional type of dense body. (From Esteve, J. C. (1970) *J. Protozool.*, 17, 23—34. Reprinted with permission of the Society of Protozoologists.)

vacuoles in this particular case. The possible interrelationships of the various acid hydrolase containing particles as exemplified by *P. caudatum* are represented diagrammatically in Fig. 3.2. It appears that the ideas expressed in this scheme are in accordance with the mainstream of thought although some protozoologists may hold different views.

The earliest evidence suggesting the fusion of primary lysosomes of metazoan cells with pinocytic vesicles came from the elegant work of Straus with a peroxidase of plant origin as a protein marker (Chapter 1, p. 12). Straus showed that following intravenous injection of rats with horseradish peroxidase the enzyme was taken up during pinocytosis by the cells of the convoluted tubules of the kidney. About 30 min. subsequent to injection the peroxidase could be detected as a *blue* reaction product, by use of the benzidine procedure for the histochemical detection of peroxidase, at or near the plasma membrane. Since these particular cells normally contained little or no endogenous peroxidase it was concluded that the injected enzyme had been taken up by pinocytosis. It was also possible in such tissues to demonstrate the simultaneous presence of pre-existing lysosomes by a standard histochemical azo dye method which gave a *red* reaction product. Moreover, simultaneous application of both procedures over a time-course study revealed the interrelationships of the lysosomes and the pinocytic vesicles (phagosomes). Observations made 6 hr after injection showed both reaction products at the same sites in the cell as indicated by the *purple* colour of the particles. These structures were termed phago-lysosomes and are equivalent to secondary lysosomes. Later in the time-course studies, two to three days after peroxidase injection, cells stained positively for acid phosphatase only; a result which was interpreted as presumptive evidence for digestion of the foreign peroxidase protein within the secondary lysosomes. These observations were reported in a long series of communications and were summarized by Straus (1967) who illustrated his findings with beautiful coloured photographs. Many later observers confirmed these relationships of the lysosomes to intracellular digestion in different tissues and with substances other than proteins, e.g. polysaccharides and fluorescently-labelled compounds of various sorts, and Straus has also been able to correlate his histochemical observations with colorimetric estimations of the content of injected peroxidase in isolated kidney fractions. These histochemical observations were confirmed at the ultrastructure level by Graham and Karnovsky (1966) using a reaction involving 3,3'-diaminobenzidine (DAB) which yields on oxidation, through a coupling reaction with peroxidase, an electron-dense polymeric product (Figs. 3.3 and 3.4).

A most elegant and impressive demonstration of the relationships between pinocytosis and the lysosome system was achieved by Maunsbach (1966), who perfused single proximal tubules of the rat kidney with [125]I-labelled rat serum albumin. Tubules were subsequently fixed by microperfusion with glutaraldehyde at various time intervals after albumin

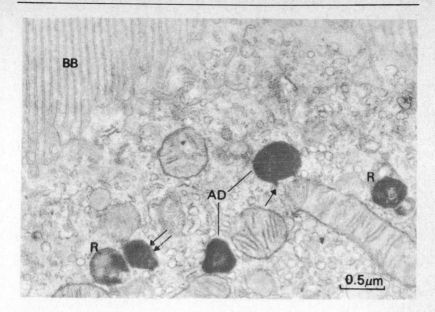

Fig. 3.3 Ultrastructural evidence for the uptake of exogenous protein into phagosomes.

Thin section of a kidney from a mouse killed 45 min. after peroxidase injection. Several absorption droplets (AD), which contain peroxidase revealed by the DAB procedure, appear to be enclosed by a single membrane. One of the absorption droplets is continuous with a tubular structure (arrow). Several bodies (R) are present, one of which (double arrow) probably contains reaction product. No reaction product is visible in the brush border (BB) or apical tubular invaginations (T). (From Graham, R. C. and Karnovsky, M. J. (1966) *J. Histochem. Cytochem.*, 14, 291–302. By courtesy of the Histochemical Society.)

administration, and ultrathin sections were examined by autoradiography in the electron microscope. No unnatural changes in ultrastructure attributable to isotopic stress were observed in tubules during the absorption process, and after 6–10 min. from the start of $^{125}$I-albumin perfusion most of the absorbed protein was located in the large apical vacuoles in the proximal tubule cells, but some was associated with small apical vacuoles suggesting that the labelled albumin was transferred to the former via the small vacuoles which probably represented pinched-off invaginations of the apical cell membrane. Thirty minutes after albumin perfusion the labelled protein was present in cytoplasmic bodies, limited by a triple-layered membrane about 90 Å thick, which were located in the middle and apical regions of the cell. After 60 min. the label was located in similar bodies in all regions of the cell, and combined autoradiography (to reveal $^{125}$I-labelled albumin) and ultrastructural histochemistry of acid

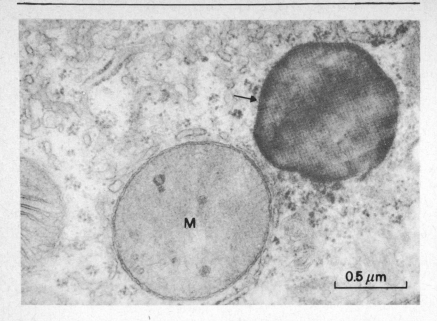

**Fig. 3.4** Ultrastructural evidence for the uptake of exogenous protein by kidney cells and its enclosure in ambilysosomes.
Mouse killed 30 min. after injection of peroxidase. The structure containing reaction product (for peroxidase) appears to contain internal membranes and to have a double limiting membrane (arrow). This structure probably represents a vacuole containing protein (peroxidase) taken in during heterophagy and membrane material accumulated by autophagy, i.e. it is an ambilysosome. (From Graham, R. C. and Karnovsky, M. J. (1966) *J. Histochem. Cytochem.*, 14, 291—302. By courtesy of the Histochemical Society.)

phosphatase showed accumulation of the albumin in lysosomes where it was believed to undergo degradation (Fig. 3.5).

The fate of solid matter following uptake by phagocytosis has been examined using markers which are electron-dense such as mercuric sulphide, thorium dioxide, gold, iron and carbon. Relationships between this process, lysosomes and intracellular digestion were clearly demonstrated by Gordon, Miller and Bensch (1965) using a gold-labelled DNA—protein complex which was rapidly phagocytosed by cultured mouse fibroblasts during which the colloidal gold component served as a marker in ultrastructural studies. This work convincingly demonstrated the relationship between Golgi-derived vesicles that conveyed hydrolases to the phagosome, which was previously devoid of such activity, and the ensuing digestive process. The multivesicular nature of the secondary lysosome was due to the continuous acquisition of Golgi-derived vesicles

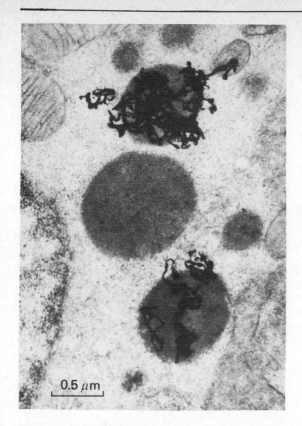

**Fig. 3.5**  Demonstration of heterophagy in rat kidney.
Electron microscope autoradiograph of a section of proximal tubule of rat kidney, which was fixed 1 hr after the absorption of a pulse of $^{125}$I-labelled albumin and which was incubated for acid phosphatase. Autoradiographic (AG) grains (small dense dots) are confined to acid phosphatase-containing bodies. This experiment proves that albumin and a lysosomal enzyme are present in the same body which is thereby established as a secondary lysosome. (From Maunsbach, A. B. (1966) *J. Ultrastruct. Res.*, 15, 197–241. By courtesy of Academic Press. Original print supplied by Professor A. B. Maunsbach.)

throughout the digestive cycle. Additionally, this work cleverly revealed the fusion of old secondary lysosomes, pre-labelled with colloidal iron prior to phagocytosis of the gold particles, with newly-formed vacuoles marked by their content of the later-administered gold. This suggested that the lysosomal enzymes may be re-used in several digestion cycles. Eventually, residual bodies became devoid of hydrolases, presumably following a process involving self-digestion by the lysosome of its own hydrolases through the action of internal proteases. The factors governing the onset of self-digestion of hydrolases within the lysosomes are unknown, although the regulation of such a process is of vital importance to the ordered functioning of cellular catabolism.

### 3.2.1 Relationships of Heterophagy to Bacterium and Virus Infections

*Bacteria.* This section could have been included in Chapter 6 which deals with lysosomes and disease, but since heterophagy plays a key role in the defence of the cell against infection and is therefore a component of the normal system I have chosen to include it here.

The association of phagocytic cells with the control of bacterium infection was noted by Metchnikoff in 1893 who observed that bacteria in tissues became surrounded by vacuoles which had acidic contents. We now know that these vacuoles acquire acid hydrolases from lysosomes along with antibacterial substances from other cell granules; thus these digestive vacuoles are secondary lysosomes. Some bacteria, e.g. tubercle bacilli and brucellae, are resistant to digestion by virtue of the composition of their cell envelopes and they may live and reproduce within the phagocytic vacuoles. In these circumstances the phagocytes, being mobile components of the circulatory system, potentially provide a most efficient means of disemminating both the pathogenic bacteria and the toxins they may produce throughout the animal body. Moreover, other types of bacteria not only survive digestion but can attack the predatory cells by means of toxins which may be released to the cytoplasm of the host cell or even cause extensive damage to surrounding tissues. In certain unusual fatal diseases the process of fusion between the phagocytic vacuole containing bacteria and the lysosomes does not occur so that the capacity to digest invaders is lost. Yet another type of response to bacterium infection is that involving production of certain toxins by the bacteria, known as streptolysins, which are thought to cause lysosome rupture followed by widespread cell disorganization and death. In this type of response it has been possible to show that release of malate dehydrogenase from mitochondria follows recognizable lysosome damage, thus encouraging the suggestion that the lysosome system may be a primary target in some bacterium infections.

Since the lysosome system is known to play a key role in the defences against bacterium infection, a number of workers have been led to believe that regulation and expression of cell immunity must also be related in some way to lysosome activity. Many attempts have been made to detect structural or functional changes in lysosomes and their hydrolases during changed immune capability but little concrete fact has emerged yet from this very active field of research, although it does seem that particular antigens must interact with macrophages *before* antibody production can take place. In this way it is envisaged that the lysosome system plays a key role in processing the antigen in a manner that specific antigenic characteristics are retained. Failure of the lysosomes to process the ingested antigen in the correct fashion or of the macrophage to transfer information regarding the nature of the antigen to potential antibody-forming cells could result in catastrophe for the organism. Those observations on the induction of pinocytosis in mouse macrophages grown in calf-serum, when large numbers of primary and secondary lysosomes are produced, provided a model system for observations on the control mechanisms regulating acid hydrolase synthesis and lysosome production which has been adapted for studying the responses of lysosomes to a number of antigens.

*Viruses.* Although viruses are structurally simple they represent the ultimate degree of specialization in parasitism. Viruses become adsorbed to the surface of the animal host cell and are taken in by phagocytosis. Once inside the cell the virus particle is uncoated and its genome penetrates that of the host where it is replicated. The replicated components are then assembled into infectious progeny particles which are released from the cell to bring about re-infection elsewhere. We have already seen that phagosomes fuse with lysosomes; such a process could result in destruction of viruses through the activities of RNase and DNase within the lysosome and this, presumably, might be the normal response whereby infection is controlled through digestion of the virus as in the case of other foreign matter. Generally, therefore, the lysosome system could cope with infections and render them abortive. Only with *Rheovirus* do the lysosomes seem to serve other than a protective role in virus disease and there have been convincing demonstrations of the uptake of this type of virus particle by the lysosomes of strain-L mouse fibroblasts where they become *uncoated*, presumably by the action of proteases. The *double-stranded* RNA genome of these particular viruses resists digestion since the RNase of mouse fibroblast lysosomes, at least, is only capable of hydrolysing single-stranded RNA. The naked RNA then passes to the cytoplasm and through its control of the host metabolism becomes replicated. So far this feature of lysosomal uncoating of viruses seems confined to *Rheovirus*.

With some other types of virus the lysosomes are not known to be involved *directly* in any part of the *replicative* cycle. In certain highly virulent viruses such as picornia (i.e. small RNA) viruses the end of the replicative cycle is accompanied by interference with the mechanisms controlling permeability of the cell. It seems that disintegration of the plasma membrane is only the final manifestation of a series of events which starts with nuclear disorganization and involves changes in lysosomal stability and culminates in cell death. There is evidence that cytotoxins may be produced by the viruses, or by the host under direction of the virus, which may cause release of acid hydrolases from the host cell lysosomes accompanied by rapid cell lysis. The use of stabilizers of lysosome membranes, e.g. hydrocortisones, may delay the effects of these toxins on lysosomes to some extent. Workers in the sphere of animal viruses believe such changes in lysosomal membrane permeability may result in cell death.

There is little information to date on the involvement of lysosomes in plant host—virus interaction although there is tenuous evidence that wound tumour virus disease (some workers consider that wound tumour may not even be caused by a virus but by a mycoplasma) may be associated with sub-cellular particles which morphologically resemble lysosomes.

### 3.2.2 Heterophagy in Plants

I have already mentioned the capacity of plant cells to carry out pinocytosis, but this process appears, as far as is known at present, to serve only a minor physiological role. Recently the uptake of DNA derived from *Escherichia coli* has been observed in tissue culture cells of sycamore, presumably by the process of endocytosis. This process of uptake of nucleic acids may be important in relation to the manner in which virus particles are taken up by plant cells, although the exact mechanism of the latter is obscure and also involves insect vectors and cell wall abrasion. In conclusion, plant cells appear to have some capacity for endocytosis but virtually nothing is known at present about its possible relationship to digestive events.

## 3.3   Autophagic Functions of Lysosomes

We have seen how lysosomes play a key role in heterophagy which forms the essential basis of nutrition in protozoa and that the process seemingly serves a mainly protective function in endothelial tissues and in the RE system of higher animals even though lysosomes are present in virtually every animal cell. This should not be taken to imply that in cells devoid of defensive capacity, or which carry out only restricted phagocytosis, the system is vestigial since it serves other key roles, prominent amongst these being participation in *autophagic* functions. This property of cells to digest portions of their own cytoplasm was known sometime before enunciation of the lysosome concept, but its relationship with lysosomes has now been demonstrated many times by histochemical, ultrastructural and biochemical techniques.

The principal method of studying cellular autophagy has been through demonstration in the electron microscope of organelles screened-off from the rest of the cytoplasm in large vesicles which are bounded by a single, or sometimes, several membranes. Reports in the literature commonly refer to the enclosure of mitochondria within these vacuoles, but most types of cytoplasmic organelles are also found, Figs. 3.6, 3.7 and 3.8. Care is necessary to distinguish truly autophagic vacuoles from other organelle-containing vacuoles which may arise in cells following uptake by hetero-

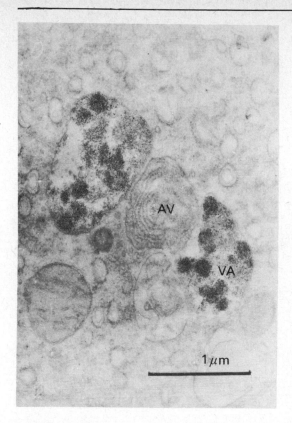

**Fig. 3.6** An autophagic vacuole (AV), containing rough endoplasmic reticulum, in close proximity to two acid phosphatase-rich vacuoles stained by an azo dye method. One of the vacuoles (VA) appears to be surrounding the autophagic vacuole. (From Bowen, I. D. (1971) *J. Microsc.*, **94**, 25–38. Original print supplied by Dr I. D. Bowen.)

phagy of cell detritus and organelles from other cells. The latter process does occur freely in scavenging cells, such as macrophages, but experimental techniques have been developed which demonstrate the rapid production of autophagic vacuoles in cells subjected to certain induction procedures, termination of which may be accompanied by a gradual disappearance of autophagic vacuoles. Progressive loss of morphological characteristics of the enclosed organelles can also be followed in time-course studies which along with the demonstrable presence of acid hydrolases strongly suggests that digestive processes are at work. From time to time published electron micrographs show vacuoles which contain materials of both exogenous and endogenous origins. Classic experiments

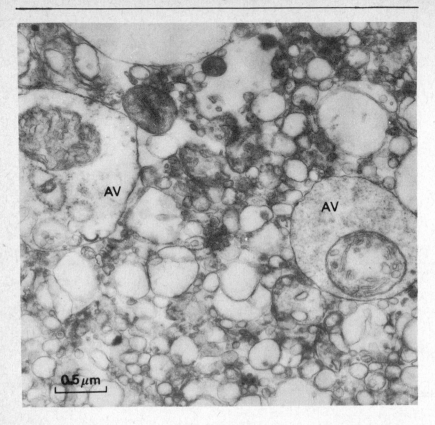

**Fig. 3.7** Thin section of the 'heavy' lysosome fraction isolated from potato sprouts showing large autophagic vacuoles (AV) containing mitochondria and other components undergoing digestion. (From Pitt, D. and Galpin, M. (1973) *Planta (Berl.)*, 109, 233–58. © Springer-Verlag, 1973.)

by Miller and Palade (1964) followed the accumulation of administered haemoglobin in mouse kidney when the proximal cells apparently segregated this compound within common vacuoles along with mitochondria and other organelles. It was also possible to demonstrate acid phosphatase in such structures and to observe lysis of the enclosed components (Fig. 3.9). Autophagy and heterophagy occurring simultaneously within the same vacuole has also been induced experimentally by injecting rats with Triton WR-1339, a detergent which accumulates in large heterophagic vacuoles which may also contain morphologically recognizable mitochondria. Thus, functionally, both autophagic bodies and combined autophagic and heterophagic vacuoles, known as *ambilysosomes*, are categorized as secondary lysosomes.

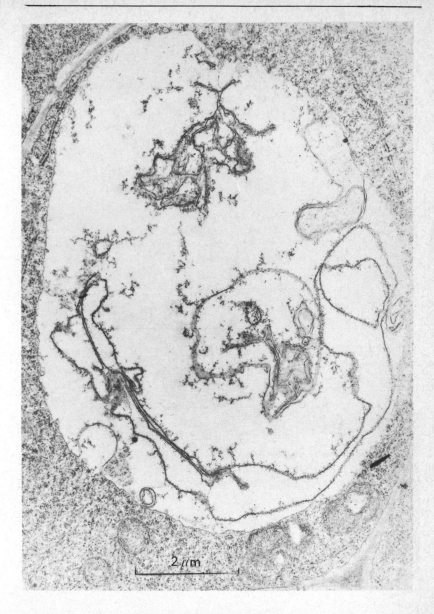

**Fig. 3.8** Electron micrograph of a section of pea root showing an autophagic vacuole. (From Hall, J. L., Flowers, T. J. and Roberts R. M. (1973) *Cell Structure and Metabolism*, Longman. Original print supplied by Dr J. L. Hall.)

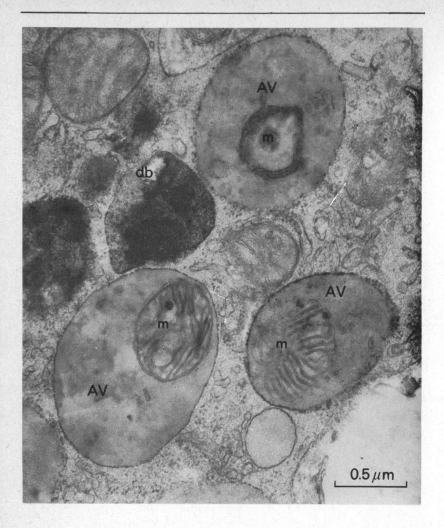

**Fig. 3.9** Autophagy and heterophagy in a cell of mouse kidney. Electron micrograph of a thin section of a kidney cell of a female mouse injected intraperitoneally with haemoglobin 2 hr previously. Note the reaction product (rp) for acid phosphatase around the bodies and digested remnants of mitochondria (m) in the interior of several autophagic vacuoles (AV). Reaction product is also associated with dense bodies (db). Thus, the autophagic vacuoles also contain haemoglobin taken in during heterophagy, i.e. they represent ambilysosomes. (From Miller, F. and Palade, G. E. (1964) *J. Cell Biol.*, 23, 519—52. By courtesy of The Rockefeller University Press. Original print supplied by Dr F. Miller.)

Extensive reports now confirm that cellular autophagy is a characteristic of normal tissues and may play a key part in the economy of the cell by participating in *turnover* of cell constituents. Some experiments in this difficult field of research have shed light on the significance of the process in re-cycling cellular commodities. Liver mitochondria appear to have a half-life of about ten days, and calculations show that if one autophagic vacuole containing a single mitochondrion were formed every 15 min. per liver cell, this could account for the observed rate of mitochondrial turnover. On the other hand however, other cytoplasmic components turn over at different rates so that simultaneous random enclosure of a variety of organelles in pieces of sequestered cytoplasm does not seem likely and some selective mechanism must be involved in the process. These data led to the proposal that only old, worn-out or defective organelles were selected for digestion, but the nature of such a selection mechanism is obscure. More recently careful experiments have been carried out in which the protein component of the peroxisomes of liver cells was radioactively labelled and the distribution of the label in sequentially isolated peroxisomes determined in time-course studies. These experiments showed that the rate of loss of radioactivity from the pulse-labelled peroxisomes occurred at a constant rate indicating that old (labelled) peroxisomes were being broken down (presumably through autophagy) at a rate similar to young (unlabelled) peroxisomes which is indicative of a *random* digestion process. The current belief, based largely on the peroxisome work, is that sequestration of organelles during autophagy is under a surprisingly crude form of control for such a vital process. Further work is needed in this important field of endeavour to determine if the process is governed other than by the law of chance.

Electron microscopy has permitted the recognition and ultrastructural characterization of the vacuole system of the cell. Even the great resolution of this technique cannot yet recognize digestive processes occurring at the level of single molecules or small molecular aggregations. Consequently, we do not know if molecular turnover is within the ambit of the lysosome system. It is known, however, that when organelles are enclosed within autophagic vacuoles considerable portions of the surrounding hyaloplasm are also engulfed and, presumably, digested along with the organelles.

### 3.3.1 Autophagy as a Response to Adverse Conditions
Cellular autophagy of the type outlined above is now established as a feature of normally functioning cells of animals, plants and protists. The process is enhanced in cells of tissues that have been deprived of essential nutrients and autophagic bodies have been found in increased numbers in long-dormant plant embryo tissues of ash (*Fraxinus*) seeds and in carbon-starved *Tetrahymena* and *Euglena* as well as in various tissues of starved

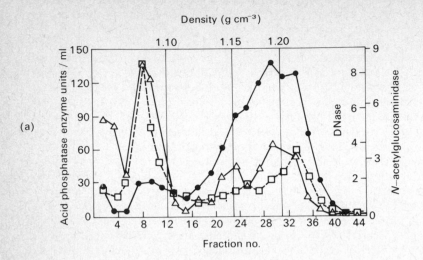

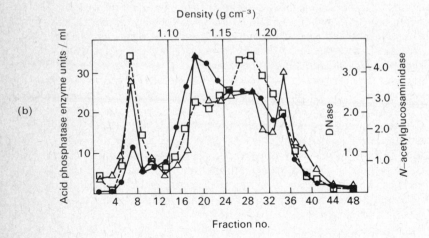

**Fig. 3.10** Changes in the density distribution of marker enzymes of lysosomes during starvation of *Tetrahymena pyriformis*.

(*a*) Shows the distribution of lysosomal enzymes after fractionation by ultracentrifugation of homogenates of one day cultures of the organism. Note two major peaks, at fractions 8 and 32.

(*b*) Shows the changed distribution of the same marker enzymes of homogenates derived from starved organisms. ●————●, acid phosphatase; △————△, deoxyribonuclease; □ ————□, *N*-acetyl-β-D-glucosaminidase.

(From Lloyd, D. *et al.* (1971) *J. gen. Microbiol.*, **65**, 209—23. By permission of Cambridge University Press.)

animals. Protozoa have been extensively studied in this respect since they are particularly amenable to biochemical investigation as growth can be precisely controlled and examinations carried out on synchronous populations. Earlier work on *Tetrahymena pyriformis* revealed ultrastructural changes on starvation which involved sequestration of mitochondria and other organelles in acid phosphatase-containing vacuoles in which digestion occurred. Starvation on organic media does not lead to changes in the specific activity of acid phosphatase but that of cathepsin (a proteolytic enzyme) increases threefold, apparently due to a loss of total cell protein rather than from a net synthesis of the enzyme in this particular system. The end-products of digestion occurred as myelin figures which accumulated in residual bodies (Levy and Elliot, 1968). Recent attempts have been made to exploit the convenient experimental systems involving protozoa in an attempt to relate the observed ultrastructural changes in development, starvation and aging to biochemical phenomena. Some results of this type of work are summarized in Fig. 3.10 which shows the changes in distribution of DNase, acid phosphatase and *N*-acetyl-glucosaminidase during *starvation* of *T. pyriformis*. These pronounced changes in density distribution of lysosomal enzymes occurring on starvation have not yet been coupled to ultrastructural observations which show increased numbers of autophagic vacuoles in the fractions containing new peaks of lysosomal marker enzymes appearing between densities 1.10 and 1.15, but this might be anticipated as a possibility. It is surprising to note that although *aging* processes are thought to involve increased autophagy no significant shifts in lysosome distribution were observed in these studies by Poole *et al.* (1971).

Starved *Euglena* cells also show increased cellular autophagy, but this system has been less extensively examined than *Tetrahymena*, although it is known that during phosphate deficiency changes occur in the relative proportions and cellular locations of two phosphatase isoenzymes possessing different substrate specificities. Transfer of the aquatic plant, *Spirodela oligorrhiza*, to phosphate-deficient conditions resulted in a fiftyfold increase in phosphatase activity of cell-free tissue extracts. This increase was accompanied by the appearance of two low molecular weight forms of the enzyme which differed in mobility on polyacrylamide gel electrophoresis from the high molecular weight form of the enzyme characteristic of plants grown under normal conditions. Marked differential changes also occurred in the ribonuclease isoenzyme complement under similar conditions. These workers (Reid and Bieleski, 1970) also found that such changes seemed to be accompanied by *de novo* synthesis of the enzymes concerned. It is tempting to suggest that these metabolic changes triggered by phosphate deficiency in duckweed (*Spirodela*) may also be associated with autophagy, but further work is needed before drawing conclusions on this point. In those instances where starvation is accompanied by increased autophagy it would seem that the latter process represents a survival mechanism whereby dispensable cell components may be hydrolysed to yield constituents which can be re-used or serve as energy sources.

### 3.3.2 Autophagy in Physiological Remodelling

The earliest observations on the occurrence of autophagy suggested that the process was a feature characteristic of tissues involved in physiological remodelling in, e.g. embryonic development, metamorphosis, physiological or induced involution. Subsequent work has confirmed these early suggestions and we shall consider some autophagic and other aspects of these processes in Chapter 5.

### 3.3.3 Autophagy as a Pathological Response

Of the many treatments which induce sub-lethal responses in cells, some are accompanied by increased numbers of autophagic vacuoles. Pathologic cellular autophagy has been experimentally induced by the administration of large doses of glucagon, a pancreatic hormone involved in glycogen breakdown in liver. Various metabolic inhibitors such as actinomycin D, an inhibitor of DNA-dependent RNA synthesis; cycloheximide, which blocks the transfer of activated amino acids from *t*-RNA to polypeptide chains; and puromycin, are all reported to cause increased autophagy. Increased cellular autophagy has also been reported in tissues subjected to various forms of radiation including X-rays, UV-radiation and γ-radiation. The literature is rich in reports of a wide range of treatments that induce focal injury in cells accompanied by increased autophagy, some of which will be mentioned in Chapter 6 but the interested reader is referred to the articles of de Duve and Wattiaux (1966), Ericsson (1969*a*) and Kerr (1973). When many of these treatments are applied in larger doses the cells usually die; observations which lead us to believe that the increased autophagy observed with milder treatments represents an attempt by the cytoplasm to screen-off an area of focal injury and prevent excessive spread of the degenerative processes which if unchecked could lead to uncontrolled autolysis culminating in cell death. Indeed Kerr (1973), suggested that the large globular structures found in the cytoplasm of sub-lethally injured liver cells arose as a result of enhanced autophagic and endocytic activity and might represent a manifestation of a regenerative process in cells that had successfully survived harmful events. However, as with all disorders in which lysosomes are involved, it is not known if injury and cell death are initially associated with lysosome malfunction or if the latter is a consequence of the former.

## 3.4 Possible Mechanisms of Formation of Autophagic Bodies

Several proposals have been made regarding the origins of the membrane(s) surrounding autophagic bodies with the merits of the various suggestions being discussed in a stimulating article by Ericsson (1969a). The GERL theory of the origin of primary lysosomes proposed by Novikoff has received strong support to date as we have already seen in Chapter 1. It is also suggested, however, that the smooth ER, containing hydrolases in flattened cisternae, envelops portions of the cytoplasm to form autophagic bodies. By this means the latter contain hydrolases from their genesis, and ultrastructural histochemistry confirms the presence of acid phosphatase associated with the enveloping membranes which by fusion could be converted to a single membrane accompanied by release of enzyme into the interior of the lysosome. Evidence favouring this concept is very strong for certain types of cell, and is particularly impressive from experiments on glucagon-induced cellular autophagy in hepatocytes when the membrane surrounding autophagic bodies was found to be rich in glucose-6-phosphatase, a specific marker of ER-derived membranes (Ericsson, 1969a). A note of discord arises however, since the single membrane of a lysosome is usually 100 Å thick whereas double membranes surrounding glucagon-induced autophagic bodies are *each* 50–60 Å thick. It is suggested, therefore, that membrane fusion and compaction occurs to give the more characteristic type of lysosome membrane which is most frequently found in these structures. Other mechanisms may also be involved about which we are currently ignorant.

Another viewpoint on the manner in which autophagic bodies acquire lysosomal enzyme activity has been promulgated by Ericsson (1969b), following critical experiments with labelled lysosomes examined by combined ultrastructural and histochemical procedures. Existing primary lysosomes of rat or mouse kidney were pre-labelled by intravenous injection of animals with electron-dense compounds, e.g. an iron–sorbitol–citric acid complex and thereby became secondary lysosomes in which acid phosphatase was also detectable. After this pre-labelling procedure the experimental animals were subjected to treatments known to induce autophagy, e.g. glucagon. Soon after induction of autophagy the autophagic vacuoles containing newly-sequestered organelles lacked the electron-dense marker. Examination of ultrathin sections of kidney tissues that had been fixed at timed intervals after glucagon administration revealed a progressive increase in the number of autophagic vacuoles containing the marker, i.e. they were ambilysosomes, and showed digestion of the enclosed organelles. These results clearly suggested that the contents of the secondary lysosomes were injected into young newly-formed autophagic vacuoles which may orginally have been devoid of lysosomal enzymes. Although these results give clear indications of the origin of and nature in which autophagic vacuoles acquire hydrolases in experimental systems it is not known if similar processes occur in autophagy in normal cells.

The present evidence, therefore, suggests that in animal cells, at least, the membranes of autophagic bodies are derived from pre-existing membranes of the ER and/or Golgi complex, probably in the region of the

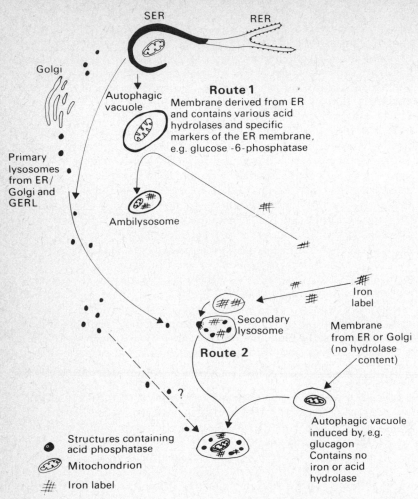

**Fig. 3.11** Suggested origins of autophagic vacuoles and their hydrolases.

*Route 1*. Membrane of autophagic vacuole (av) derived from SER which also contributes acid phosphatase. Specific markers of ER membrane material, e.g. glucose-6-phosphatase are also present in membranes of autophagic vacuoles which tends to confirm the existence of this route. Phagosomes may fuse with the av to give an ambilysosome.

*Route 2*. Iron-labelled secondary lysosomes contain acid phosphatase. Induced autophagic vacuoles (e.g. by glucagon injection) initially contain no electron-dense iron label. At longer intervals after av-induction these vacuoles contained *iron* and *hydrolase* and were recognized as ambilysosomes. Those autophagic vacuoles produced by this route gain hydrolases from *secondary lysosomes* although it is possible that they may do so via fusion with primary lysosomes on some occasions.

GERL, and that lysosomal enzyme activity may be derived directly from the enveloping membrane or its cisterna. Alternatively the ER-derived membrane may be devoid of enzymes in some cell types and hydrolase activity is acquired, by fusion, from pre-existing secondary lysosomes. The possibilities of other origins cannot be excluded in the present fluid state of knowledge and current suggestions for alternatives include the postulated enclosure of cytoplasmic material by the wrapping around of the area by a lysosome which has adopted a flattened sac arrangement; or by the accumulation within an area destined for focal degradation of numerous primary lysosomes (either Golgi or ER-derived) which then fuse to form a continuous surrounding membrane. Figure 3.11 summarizes some possible mechanisms of autophagy.

Our ideas relating to the origin of autophagic vacuoles in the protists and plants are nebulous at the present time. Proposed mechanisms are in line with those suggested by Novikoff, but the sequence may differ in detail as seen in the work on *Paramecium* mentioned earlier (p. 57, Fig. 3.2). Detailed work by Brandes *et al.* (1964) on carbon-starved *Euglena* strongly suggests the participation of the Golgi apparatus in the formation of autophagic vacuoles and as the source of the acid phosphatase associated with these structures. The origin of the extensive autophagic vacuolar system in plant cells is more obscure. Matile (1969) considers that vacuoles incorporating cytoplasmic material may arise following invagination of the tonoplast resulting in an intravacuolar vesicle. This system would not account for the images seen in many of the published electron micrographs of autophagic vacuoles, and it is apparent that our understanding awaits further work in the plant field. Recent work by Coulomb and Coulomb (1971) suggests that acid phosphatase-rich dense bodies (primary lysosomes) arise from the Golgi apparatus and fuse with the autophagic vacuoles in a way similar to that proposed by Ericsson for certain animal cells, however, further work would be required to confirm this view and also to determine if ER enclosure occurs.

## 3.5 Other Lysosomal Functions Related to Intracellular Digestion

As a consequence of the process of intracellular digestion undigested residues may accumulate in the cell within *residual bodies*. These may contain undigested or indigestible materials which can remain indefinitely inside the cell and lead to enlargement of the lysosome system and ultimate interference with the normal functioning of the cell. Such accumulations may occur naturally in some tissues as instanced by the increased lipofuscin content in aging human nerve, heart and liver. In some cases the lysosomes may accumulate *abnormal* materials generated by the cells because the *normal* complement of lysosomal enzymes cannot

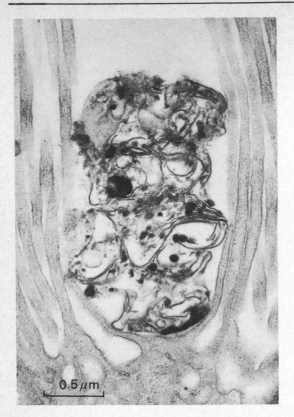

**Fig. 3.12**  Electron micrograph of a thin section of rat kidney showing an extra-cellular body between the brush border and the lumen of the proximal tubule. The original work also showed that such bodies contained acid phosphatase. These extruded bodies represent former residual bodies which have been passed from cells by reversed pinocytosis, i.e. they represent a means of cellular defaecation. (From Maunsbach, A. B. (1966) *J. Ultrastruct. Res.*, 16, 197–238. Courtesy of Academic Press Inc. Original print supplied by Professor A. B. Maunsbach.)

hydrolyse the compound. A number of *inborn genetic disorders* may result in the failure of a cell to metabolize a normal cellular commodity which then accumulates within the cell as in the case of certain *storage disorders* which we shall consider in more detail in Chapter 6.

It is possible for a few cell types to discharge to the outside materials which occur in lysosomes. In Chapter 4 we shall consider the *secretory* function of lysosomes and their role in *extracellular digestion*. Additionally, some animal cells have the rare capacity to discharge waste materials through *exocytic* discharge in the process of *defaecation* similar to that

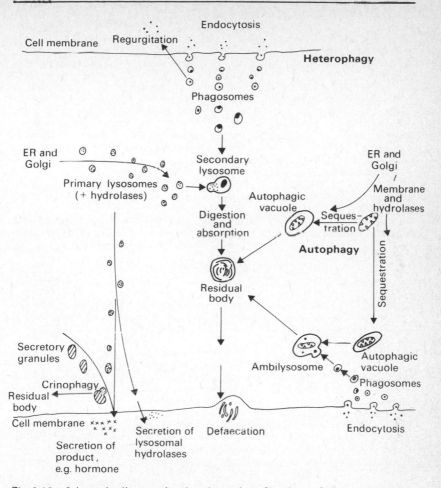

**Fig. 3.13** Schematic diagram showing the various functions of lysosomes and the relationships between heterophagy and autophagy.

found in protozoa. The discharge of myelinated material by kidney cells is shown in Fig. 3.12. Finally, it is known that certain cells may be rich in phagosomes yet have relatively few lysosomes. In such conditions it may be that the lysosome system is involved in the *transport* of materials within cells and across cell barriers. The recent increased knowledge of one function of microtubules in intracellular transport could suggest a means of achieving such movements via the lysosome system.

Some of the known interrelationships and functions of the lysosome system are summarized in Fig. 3.13.

## Suggestions for further reading

CHAPMAN-ANDRESEN, C. (1962) Studies in pinocytosis in amoebae, *Compt. Rend. Trav. Lab.*, *Carlsberg*, 33, 73—264.

de DUVE, C. and WATTIAUX, R. (1966) Functions of lysosomes, *Ann. Rev. Physiol.*, 28, 435—92.

ERICSSON, J. L. E. (1969*a*) Mechanism of cellular autophagy, in *Frontiers of Biology*, 14B. *Lysosomes in Biology and Pathology*, vol. 2, ch. 12, pp. 346—9, eds. J. T. Dingle and H. B. Fell. North-Holland Publishing Co., Amsterdam and London.

ERICSSON, J. L. E. (1969*b*) Studies on induced cellular autophagy. J. Electron-microscopy of cells with *in vivo* labelled lysosomes, *Exp. Cell. Res.*, 55, 95—106.

ESTEVE, J. C. (1970) Distribution of acid phosphatase in *Paramecium caudatum*: Its relations with the process of digestion, *J. Protozool.*, 17, 23—34.

GORDON, G. B., MILLER, L. R. and BENSCH, K. G. (1965) Studies on the intracellular digestive process in mammalian tissue culture cells, *J. Cell Biol.*, 25, 41—56.

GRAHAM, R. C. and KARNOVSKY, M. J. (1966) The early stages of absorption of injected horseradish peroxidase in the proximal tubules of mouse kidney: ultrastructural cytochemistry by a new technique, *J. Histochem. Cytochem.*, 14, 291—302.

HOLTER, H. (1959) Pinocytosis, *Int. Rev. Cytol.*, 8, 481—504.

HOLTER, H. (1961) *Proc. Fifth Int. Congress. Biochem.*, Moscow, vol. 2, p. 248. Pergamon Press, New York.

KERR, J. F. R. (1973) Some lysosome functions in liver cells reacting to sublethal injury, in *Frontiers of Biology*, 29. *Lysosomes in Biology and Pathology*, vol. 3, ch. 14, pp. 365—94, ed. J. T. Dingle. North-Holland Publishing Co., Amsterdam and London.

LEVY, M. R. and ELLIOTT, A. M. (1968) Biochemical and ultrastructural changes in *Tetrahymena pyriformis* during starvation, *J. Protozool.*, 15, 208—22.

MATILE, P. (1969) Plant Lysosomes, in *Frontiers of Biology*, 14A. *Lysosomes in Biology and Pathology*, vol. 1, ch. 15, pp. 406—30, eds. J. T. Dingle and H. B. Fell. North-Holland Publishing Co., Amsterdam and London.

MAHLBERG, P. (1972) Localization of neutral red in lysosome structures in hair cells of *Tradescantia virginiana*, *Can. J. Bot.*, 50, 857—9.

MAUNSBACH, A. B. (1966*a*) Absorption of [125]I-labelled homologous albumin by rat kidney proximal tubules cells: a study of microperfused single proximal tubules by electron microscopic autoradiography and histochemistry, *J. Ultrastruct. Res.*, 15, 197—241.

MAUNSBACH, A. B. (1966*b*) Observations on the ultrastructure and acid phosphatase activity of the cytoplasmic bodies in rat kidney proximal tubule cells, *J. Ultrastruct. Res.*, 16, 197—238.

MILLER, F. and PALADE, G. E. (1964) Lytic activities in renal protein absorption droplets, *J. Cell Biol.*, 23, 519—52.

REID, M. S. and BIELESKI, R. L. (1970) Changes in phosphatase activity in phosphorus-deficient *Spirodela*, *Planta* (*Berl.*), 94, 273—81.

STRAUS, W. (1967) Lysosomes, phagosomes and related particles, in *Enzyme Cytology*, pp. 239—319. ed. D. B. Roodyn. Academic Press, London and New York.

WATTIAUX, R. (1969) Biochemistry and functions of lysosomes, in *Handbook of Molecular Cytology*, pp. 1159—78, ed. A. Lima-de-Faria. North-Holland Publishing Co., Amsterdam and London.

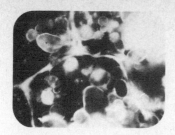

Chapter 4

# Physiological Functions of Lysosomes: Role of Lysosomes in Secretion and Extracellular Digestion

In recent years a number of reports of studies on various cell types from a wide range of plants and animals have suggested that the activity of acid hydrolases in some secretory cells may be linked with the formation of the secretory product. Particularly strong evidence has been provided by Smith and Farquhar (1966) who found intracellular lytic activity, as indicated by ultrastructural histochemical localization of acid phosphatase, was a major feature of secretory cells of the anterior lobe of the rat pituitary and was involved in both *secretion* and *modulation* of secretory function. It is also known that lysosomal enzymes are themselves secreted by *reversed pinocytosis* (exocytosis), although the mechanism by which this occurs is obscure and involves fusion of lysosomes with the plasma membrane. In the external medium the secreted lysosome enzymes may be involved in the normal physiological function of digestion of the extra-cellular matrix of connective tissues. This process may in turn be related to the intracellular digestive system following uptake of the breakdown products by endocytosis. The process of extracellular digestion may be exaggerated in pathological conditions.

## 4.1 Role of Lysosomes in Hormone Secretion

### 4.1.1 Secretion of Thyroid Hormones
Thyroxine and tri-iodothyronine occur in the thyroid gland linked coval-ently in a high molecular weight protein called thyroglobulin which is usually stored as a colloid in the lumen of the gland. The secreted thyroid hormones occurring in the blood are *not* in association with the protein

moiety. Accordingly, the thyroglobulin molecule must undergo hydrolysis to release the thyroid hormones. Although a sufficient complement of the necessary hydrolases (proteases and peptidases) is found in the thyroid gland these are located in the secretory cells and are spatially separated from the thyroglobulin of the lumen. Before secretion can occur it is necessary for some mechanism to prevail which brings together the lytic enzymes and the thyroglobulin.

The thyroid gland is especially amenable to experimentation since it can be rendered inactive by thyroxine feeding or stimulated by injecting animals, e.g. rats, with thyroid stimulating hormone (TSH) from the pituitary body. Stimulation by TSH results in marked morphological changes in the epithelial cells culminating in the appearance of colloid droplets in the cell, acquired from the lumen, accompanied by the gradual disappearance of colloid from the lumens of the follicles. Newly-formed colloid droplets are free from acid phosphatase activity as determined by ultrastructural histochemistry, but within 5–10 min. of their appearance within the cell they acquire detectable hydrolase activity. It is presumed that hydrolysis of thyroglobulin occurs soon after fusion of lysosomes with colloid droplets. Some evidence exists that dense granules (lysosomes) are regenerated from the body which results from the initial fusion process, i.e. from the secondary lysosome, and further participate in a cyclic process of repeated digestive bouts. During thyroid stimulation the thyroxine concentration increases in the blood circulating from the thyroid gland, presumably through hormone release in exocytosis (Fig. 4.1).

The phenomenon of lysosome-mediated thyroid hormone secretion is the best known example of direct lysosomal involvement in secretory processes. A similar mechanism could apply to other hormones, but little information is available at the present time. However, limited data are available concerning the possible link between lysosomes and another endocrine tissue, namely, the adrenal medulla. This tissue produces and stores the catecholamine hormones, adrenaline and noradrenaline, in membrane-bounded chromaffin granules in special cells known as chromaffin cells. Secretion of the hormones from the cell is achieved following the fusion of membranes of the granules with the cell membrane, but the empty remains of chromaffin granules stay within the cell and eventually become detached from the plasma membrane. It is believed that although the lysosomes may not be involved directly in the secretion process in that manner outlined for thyroid hormones, they may be associated with autophagic digestion of residual membrane materials of the empty chromaffin granule.

### 4.1.2 Lysosomes in the Regulation of Hormone Secretion
The pituitary gland holds a position of central importance in the functioning of the endocrine system. Under the influence of the hypothalamus its

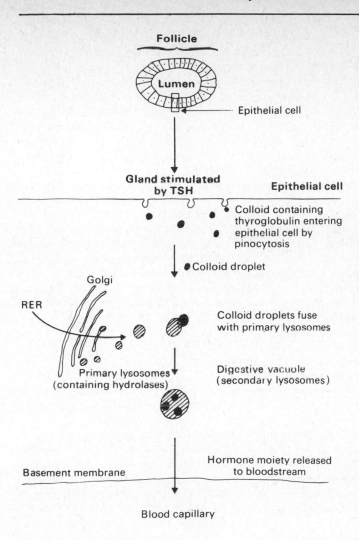

**Fig. 4.1** The thyroid gland contains several cell types with the functional unit of the gland being the follicle. Follicles are more or less spherical entities consisting of a layer of epithelial cells surrounding a lumen. The lumen contains a colloid which incorporates most of the iodine of the gland in the form of stored iodoprotein, particularly thyroglobulin. The follicles are surrounded by a basement membrane and are served with capillaries. Thyroid epithelial cells contain conspicuous lysosomes rich in lysosomal enzymes. It is believed that the hormone moiety is split from the stored thyroglobulin by controlled digestion within the secondary lysosomes with the hormone then being released to the capillaries.

anterior lobe is responsible for producing six hormones including follicle-stimulating hormone, luteinizing hormone and lactogenic hormone (prolactin) which all act directly on particular sex glands and thereby regulate, in females, ovulation, pregnancy and lactation. In addition the anterior pituitary gland produces adrenocortical trophic hormone (ACTH) which promotes formation of the steroid hormones of the adrenal cortex. The growth-regulating hormone (somatotrophic hormone or STH), and thyrotrophin (TSH) are also produced by the anterior lobe of the pituitary. It is believed that each hormone is produced in a special cell type, e.g. lactogenic hormone in the mammotrophic cells, growth hormone in somatrophs and TSH in the thyrotrophs.

Some progress has been made in cell fractionation of the anterior pituitary whereby secretory granules have been isolated and recognized as the cellular storage sites of the particular hormones. Of the various types of secretory cell found in the anterior lobe of the pituitary the mammo-troph has proved most suitable for study because of its relatively large size and the convenient manner in which its secretory functions may be manipulated for experimental purposes. Stimulated mammotrophic cells may be obtained from the pituitary bodies of lactating animals. Suppression of prolactin secretion can be conveniently achieved by removal of the suckling young since the suckling process appears to act through the hypothalamus by suppressing the release of prolactin inhibition factor.

The mammotrophs of lactating animals with suckling young show many secretory granules. Steps in the synthesis and intracellular transport of secretory granules seem to follow the classical pattern established by Palade's group for zymogen in the exocrine cells of the pancreas, in which protein, synthesized at the ribosomes, is transferred to the cisternae of the ER and passes in vesicular form to the Golgi apparatus where it is packaged into secretory granules. Within 24 hr of suppressing secretion of prolactin by removal of the suckling young, some of the secretory granules may be detected within lysosomes by use of appropriate techniques of ultrastructural histochemistry. The sequential details of the process of incorporation and digestion may be followed at the ultrastructural level by means of time-course experiments in which it can be observed that the decrease in the number of free secretory granules is accompanied by increased incorporation of granules into lysosomes. The sequence also reveals progressive degradation of sequestered granules to leave myelin figures, membrane debris and lipid materials. Concurrent with this chain of events there is an increase in the incidence of autophagic vacuoles in the mammotrophic cells which seem to be primarily involved in dismantling the protein manufacturing and secreting apparatus of the rough ER rather than in the digestion of the granules. The series of events as recognized in the rat pituitary is represented schematically in Fig. 4.2.

Other secretory cells of the anterior pituitary have been examined in less detail since they are not as suited to experimental manipulation as

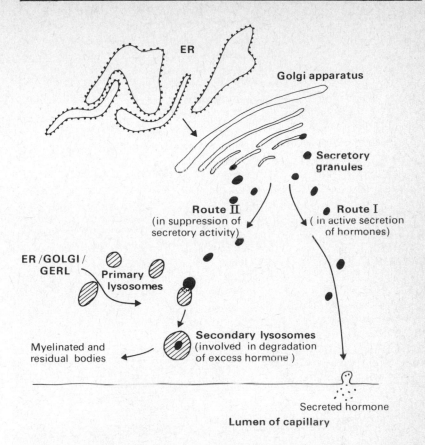

**Fig. 4.2** Outline scheme of the possible role of lysosomes in the modulation of hormone secretion in mammotrophs of the anterior pituitary gland. It is believed that mammotrophic hormone is synthesized on the ribosomes of the RER and is packaged into secretory granules by passage through the Golgi (cf. zymogen in the exocrine cells of the pancreas). When cells maintain a high level of secretion Route I is followed. During reduced secretory activity excess of stored hormone is degraded after fusion with lysosomes (Route II) in the process of crinophagy.

the mammotrophs. However, the information that is available indicates that secretion by the somatrophs and thyrotrophs also involves lysosome-mediated hydrolysis. It also seems that the lysosome systems of these pituitary cells are likewise involved in the disposal of excess secretory products, and there is now enough information available to suggest that such secretory processes are under feedback control regulation from the target endocrine organs.

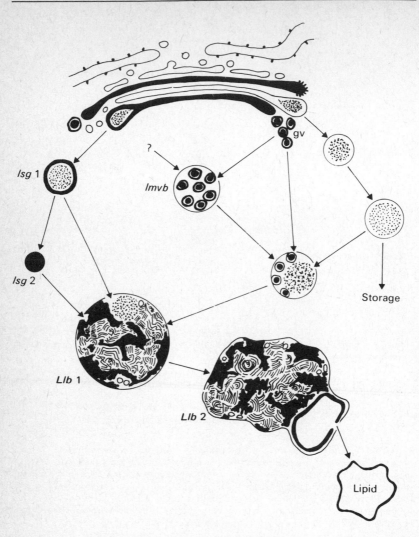

**Fig. 4.3**  A schematic outline of the possible intracellular routes taken by newly-packaged secretory product (stippled) in the prolactin cell of *Poecilia latipinna* returned to sea-water for 72 hr. It is believed that acid phosphatase (coloured black) present in a proportion of the Golgi cisternae becomes included within a number (*lsg* 1) of the newly-formed secretory granules and Golgi vesicles (gv). The acid phosphatase-positive developing granules are either degraded immediately (*lsg* 2) or coalesce with larger lytic bodies (*Llb* 1). Newly-formed secretory granules lacking acid phosphatase may mature and remain in the cytoplasm or become degraded either as a result of fusing with multivesicular bodies (*lmvb*) or as a result of free Golgi vesicles containing acid phosphatase, becoming incorporated within them. As a result of this degradative activity and by fusing with other similar structures, the

multivesicular lytic bodies containing secretory product become transformed into large lytic bodies (*Llb* 1) which in turn are the precursors of the larger residual bodies (*Llb* 2). The possibility that the discrete lipid masses observed within the residual bodies are released to become free within the cytoplasm is also indicated. (From Hopkins, C. R. (1969) *Tissue and Cell*, 1, 653—71. With permission of Oliver and Boyd Ltd.)

---

A further example of the role of acid phosphatase-positive bodies in endocrine regulation is seen in the removal of surplus secretory granules of the prolactin cell of the teleost fish pituitary. When *Poecilia latipinna* is acclimatized to sea-water the prolactin cells are fairly inactive. These cells are stimulated to synthesize and release secretory product on transfer of the fish to fresh-water and their subsequent return to sea-water is accompanied by a decline in secretory activity of the prolactin cells. Changing the environmental conditions provides a convenient method for experimentally examining the secretory activity of the prolactin cells. Ultrastructural studies on the teleost pituitary (Hopkins, 1969) indicated that surplus secretory product in the prolactin cells of fish that had been transferred to fresh-water was degraded if the animals were re-introduced to sea-water when storage granules containing secretory product fused with lytic bodies, rich in acid phosphatase, that had arisen from the Golgi apparatus. Thus, it is possible to recognize in the teleost pituitary control of osmoregulation some features in common with the observations of Smith and Farquhar (1966) on the role of lysosomes in secretory regulation in the mammotrophic cells of mammals. Figure 4.3 summarizes the postulated sequence of events in this facet of osmoregulation in *Poecilia*.

## 4.2 Lysosomes in Extracellular Digestion

### 4.2.1 In Connective Tissue Catabolism
Here we shall consider the normal process of extrusion of lysosomal enzymes which participate in the hydrolysis of extracellular materials such as connective tissues. We shall not deal at present with the widespread uncoordinated acute breakdown which occurs in these tissues during some disease situations since we shall mention these in Chapter 6. Much of the work has, of necessity, taken the form of tissue culture studies owing to the biochemical problems associated with following the process in intact organisms.

The earliest observations were carried out in the Strangeways Laboratory, at Cambridge, by Dame Honor Fell's group who showed, several years before the lysosome concept was established, that when limb-bone rudiments from chick embryos were grown in culture on clotted plasma in the presence of excess vitamin A the cartilaginous matrix was broken down and the clot on which they grew was liquefied. When the lysosome concept was enunciated it was realized that secretion of a lysosomal acid protease could be involved. Stimulation of secretion of enzyme by vitamin A was considered a result of the labilization of the lysosomal membrane which was accompanied by *de novo* synthesis of this enzyme. Intensive work by Dingle and his colleagues at the Strangeways Laboratory has resulted in the isolation and purification of a specific protease, cathepsin D, from chicken liver lysosomes. Perhaps at this point I should explain that the activity of an enzyme preparation may be reduced in a dose-dependent manner within particular limits, by permitting the enzyme to complex with its specific antibody. Immunological inhibition of lysosome function was achieved following the uptake of antibodies to soluble components of rat liver lysosomes by rat fibroblasts (Tulkens, Trouet and van Hoof, 1970). The Cambridge group has extended this concept and raised in rabbits a specific antiserum to purified cathepsin D which inhibited activity of several molecular forms of cathepsin D from a number of tissues without impairing the activity of other lysosome enzymes. When limb-bone cartilage was pre-treated with the specific antiserum to cathepsin D a dose-dependent inhibition of autolytic degradation of the cartilage culture was found. These results clearly suggested that in the particular tissues examined cathepsin D was largely responsible for the hydrolysis of extracellular matrix (Dingle *et al.*, 1971). In addition the technique provides a bonus in that when the antiserum is conjugated with a fluorescent dye the sites of antigen—antibody complexes (i.e. sites of enzyme activity) in tissues may be detected by the fluorescence imparted to the complex by the fluorochrome. In this instance cathepsin D was located in tissue sections within particles having characteristics of lysosomes. These studies may provide the basis for exciting possibilities for the control of pathological conditions arising from excessive extracellular erosion of connective tissue matrix by lysosome enzymes.

Although acid hydrolases are found in many body fluids of normal animals, there is no irrefutable evidence that they are all derived from lysosomes. Thus, our knowledge of extracellular matrix digestion is based on organ culture studies which could reflect abnormalities of such systems. However, in tissue culture it is clear that several enzymes may be released from lysosomes and could account for much of the degradation occurring in some connective tissues *in vivo*. On the other hand collagen is a major component of connective tissue and there is relatively little evidence that collagenase is located in lysosomes although it does appear fairly early in the media of organ cultures undergoing resorption. Currently, apart from the example of cathepsin D in certain tissues, the situation is unclear and further research is necessary before we can understand the general nature of resorption processes.

The enzymes of lysosomes appear to be involved in a number of situations where natural physiological turnover of matrix occurs. For instance there is evidence of increased activity of lysosomes in the tails of tadpoles during metamorphosis and also in post-partum involution of the uterus of the rat. These are examples which I have chosen to deal with under the general heading of physiological remodelling during development in Chapter 5. However, I would like to consider some aspects of bone resorption which represents a specific instance of extracellular matrix degradation which has been examined in considerable detail. Bone is a complex tissue consisting of several cell types associated with a matrix composed of collagen fibres and minerals. The cell types include *osteoblasts*, which are responsible for the formation of the calcified component of bone, and the *osteoclasts*, multinucleate cells involved in the breakdown of the calcified bone matrix. A third type of cell, the *osteocyte*, is derived from the osteoblasts and gradually becomes embedded within the bone in lacunae. In normal circumstances the continuous process of bone resorption is balanced by deposition of bone substance. Remodelling of bone shape also occurs and involves an appropriate balance between bone breakdown and synthesis.

The physical properties of bone have made this tissue particularly difficult to examine by ultrastructural, histochemical and biochemical techniques. Nevertheless, appreciable progress has been achieved through the use of bone from very young animals and that grown in organ culture. Biochemical studies on those bone cells which do permit analysis have shown several particulate hydrolases which demonstrate latency and are activated by those classical treatments pertaining to lysosomes. However, evidence for the lysosomal location of collagenases in bone is as slight as that for other connective tissues, and it may be that this enzyme is secreted immediately following its synthesis rather than passing through a phase of association with lysosomes. Ultrastructural characterization of lysosomes in bone cells has been hampered owing to technical problems associated with obtaining thin sections, but it has been shown that osteoclasts are especially rich in lysosomal hydrolases and this fact coupled with their known historical role in resorption has prompted fairly intensive study of these cells in relation to bone erosion. Work carried out with bone in organ culture revealed that osteoclasts were the main agents of bone resorption although other bone cells may play a subsidiary role in the process. Where ultrastructural studies have been done they revealed that the typical osteoclast has large numbers of vacuoles and vesicles associated with the ruffled border which represents a much-folded plasma membrane that is closely applied to the area of bone undergoing resorption. The many vesicles in the cytoplasm of the osteoclast represent primary lysosomes, containing acid hydrolases, which are secreted into the immediate extracellular zone of matrix digestion. It is also known that solid materials in various states of degradation are taken into the cell by phagocytosis, and within these heterophagic bodies digestion is completed and may involve the addition of further hydrolases by fusion with primary lysosomes. Limited work that has been done suggests that the continuous

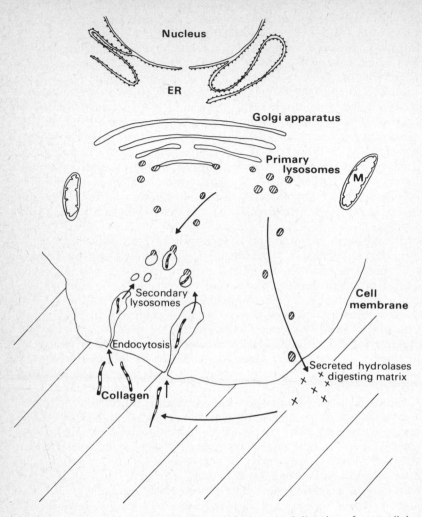

**Fig. 4.4**  A diagrammatic representation of the 'two-way' digestion of extracellular matrix, e.g. bone, by lysosomal enzymes. Hydrolases are secreted to the matrix through the mediation of primary lysosomes. Partial degradation of matrix material occurs outside the cell with both final and incomplete products of digestion being absorbed by the cell in endocytosis. Digestion of partially degraded components is completed in secondary lysosomes.

secretion of enzymes to the exterior of the cell is accompanied by *de novo* enzyme synthesis to compensate for the perpetual drain of hydrolases to the outside. Thus, digestion in the immediate microenvironment of the osteoclast membrane is accompanied by simultaneous exocytosis and phagocytosis in what is envisaged as a *'two-stage'* process involving both *extracellular* and *intracellular* digestion for the complete degradation of the extracellular matrix. Resorption is also accompanied by simultaneous demineralization of the bone as well as the degradation of the organic component, but the mechanisms involved in the former process are little understood at present. The case for implicating lysosomes in bone resorption is strengthened by the observation that whilst the process is enhanced *in vitro* by the presence of the labilizer of membranes of lysosomes, vitamin A, such stimulation may be suppressed and reversed by those compounds which stabilize lysosome membranes, e.g. hydrocortisone. A schematic outline of the probable sequence of events in bone resorption is given in Fig. 4.4. It will be noticed that bone resorption has many features in common with the similar 'two-stage' process believed to operate in the extracellular degradation of cartilage (reviewed by Dingle, 1969). In both processes it is likely that the rate and extent of degradation achieved depends upon the balance between the rates of hydrolysis and synthesis which prevail in the tissues and must, therefore, be under the control of very precise regulatory mechanisms.

## 4.3  Extracellular Secretion of Acid Hydrolases by Fungi

Perhaps the most striking example of secretion of extracellular enzymes by living organisms occurs in the bacteria and fungi which demonstrate *heterotrophy*. These organisms have two major ways of obtaining nutrients, through *saprophytism* or *parasitism*. Saprophytic fungi colonize and digest organic remains and absorb soluble substances from such material. On occasions colonization and damage may occur to living organisms when the host is said to be diseased and the parasite is termed a *pathogen*. Pathogenic fungi are grouped according to their ecological and nutritional status and include *necrotrophic* fungi which show varying degrees of parasitism and are often opportunist saprophytes, *biotrophic* forms which are obligately parasitic, and some which form *mutualistic symbiotic associations*. Since bacteria do not contain a recognizable lysosome system we shall not consider them further. All fungi are heterotrophic for carbon and some species may also have special requirements for nitrogenous compounds and vitamins. Essential nutrients may occur in a simple form in the environment from where they are absorbed directly by the mycelium. Frequently, however, complex substrates are utilized through the facility of fungi to secrete extracellular enzymes which degrade the substrate to simpler, usually soluble, forms which are taken in by the cells, e.g. carbohydrates are usually absorbed as hexoses, mainly glucose. Although fungi are able to utilize a wide range of complex organic compounds as sources of carbon, the actual availability of the particular energy source varies within the changing environment. A degree of sophistication exists in these organisms to cope with such changes in substrate availability through the facility to synthesize appropriate enzymes in response to a specific signal from the environment. In this way *inducible enzymes* are produced *de novo*, and the enzymic constitution of the mycelium may change dramatically in response to alterations of substrate availability in the environment. Many of these enzymes are secreted and degrade the substrates in the habitat. Such a system is highly efficient and combines great nutritional flexibility with energy conservation since only those enzymes which are necessary are synthesized. The process has some features in common with the inducible enzyme systems so extensively examined in bacteria, but it is not known if the regulatory mechanisms controlling this synthesis are the same as those so spectacularly demonstrated by Monod's group for procaryotic organisms. Although enzyme induction phenomena are common in the fungi, this group of organisms also has the capacity to synthesize, and in many cases to secrete, a battery of enzymes irrespective of the substrate which is immediately available. Such enzymes, which are produced at a more or less constant rate, are known as *constitutive* enzymes. Whichever mechanism prevails the net result is the same in that the enzymes produced, and either retained intracellularly or secreted, degrade substrates of varying degrees of complexity. In those cases where extracellular enzymes are involved the breakdown products are usually simple organic molecules which are then transported into the cells. The mechanisms involved in uptake by fungi of substances from the environment are little understood. Movement of a solute into these cells may occur through *passive transport* involving diffusion along a concentration gradient. However, there is some considerable degree of selectivity in uptake which can also occur against a concentration difference. Such selective uptake often demonstrates an energy requirement and is known as *active transport*. It is now believed that some form of carrier mechanism is involved since the kinetics of the process are reminiscent of those relating to enzymes. It may be that permeases are involved in a manner analogous to that in bacteria, but further work is necessary before any definite statements can be made. It is also likely that active transport may take place through the medium of pinocytosis, perhaps at the ultrastructural level, but although the process is well established in animals and is known to occur in plants Holter, in classical studies on pinocytosis, failed to find any evidence of this phenomenon in yeasts, although fungus spores are able to accumulate dyes, fungitoxicants and other large molecules by a process resembling pinocytosis.

We have already summarized the evidence for the existence of lysosomes in fungi (Chapter 2, p. 28). Under appropriate conditions enzymes with lysosome locations are secreted and appear in quantity in the external medium. A fairly substantial literature records the extracellular production of a number of these enzymes which include acid phosphatase, esterases, ribonucleases, proteases and peptidases. With appropriate supplementation of the medium, e.g. addition of RNA or organic phosphates, secretion of ribonucleases and phosphatases, respectively, may be dramatically increased from the previously low intracellular level which prevails in the absence of the amendments. In some cases a degree of inducibility is displayed, but frequently the enzymes are produced constitutively and secreted in an apparently non-selective fashion to the exterior. Such is the capacity for the constitutive secretion of ribonucleases by *Neurospora crassa* and *Aspergillus oryzae* that these organisms form the basis for the commercial production of these enzymes, particularly in Japan. In some fungi, however, certain lysosomal enzymes are confined to intracellular sites following induction. Beta-galactosidase, which is located in lysosomes in a number of fungi (Pitt, 1968), accumulates at intracellular sites in the mycelium of *Verticillium albo-atrum* grown on lactose as a source of carbon. The specific activity of this enzyme within the mycelium increases 40–200 fold in the presence of lactose and is under the control of lactose induction and catabolite repression. In addition interesting work is in progress on the location of induced a-L-arabinofuranosidase of *Sclerotinia fructigena* (Hislop *et al.* 1974). This enzyme, along with secreted pectolytic enzymes is responsible for the degradation of cell wall polymers based on araban and pectic substances during rotting of apple fruits caused by this pathogen. The a-L-arabinofuranosidase has been located by ultrastructural histochemistry at particulate sites, similar to those containing acid phosphatase, in the cells of *Sclerotinia fructigena*. If these observations are extended to other enzymes of known lysosomal locations it would permit confirmation of the speculative ideas that lysosomes of both the host and the pathogen may play significant roles in plant disease. Of special relevance not only to plant pathology but to the lysosome concept in general are the important recent observations of Grove, Bracker and Morré (1968) on the origin of secretory vesicles in fungi. These workers were able to show by electron microscopy that the membrane of the Golgi saccules widens progressively and stains more intensely with barium permanganate from its 'forming' (concave) face to the secretory (convex) face in *Pythium ultimum* (Fig. 4.5). The opinion was expressed that the major function of the apparatus in fungi is to elaborate secretory vesicles, the membrane of which can fuse specifically with the plasma membrane. Since fungi are particularly amenable to experimentation and demonstrate a mode of life so dependent upon secretion they provide excellent material for elucidating the nature of membrane transformation through the Golgi body and the origin of primary lysosomes and their role in extracellular secretion. Although it is believed that the lysosomes of fungi serve, in addition to their role in secretion, an intracellular digestive function involving autophagy, there is no evidence to date that they participate in heterophagic phenomena and in this way their mode of heterotrophic nutrition differs from the 'two-way' digestion of extracellular matrix

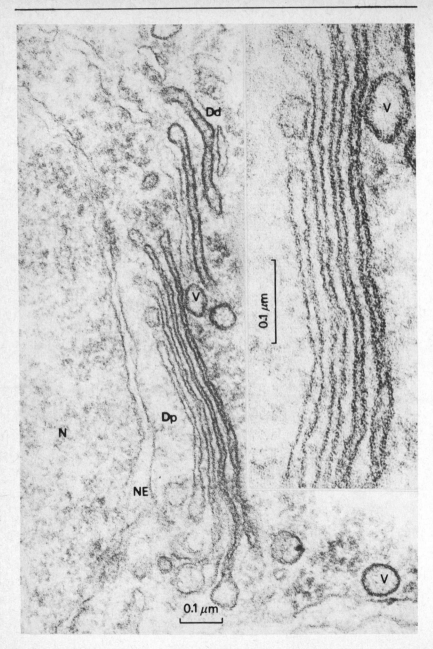

**Fig. 4.5**   Origin of secretory bodies in fungi. A dictyosome and associated secretory vesicles (V) adjacent to a nucleus (N). The membrane of the cisterna at the proximal

pole of the dictyosome (Dp), is similar to that of the nuclear envelope (NE). The membranes of each successive cisterna stain more intensely and appear thicker towards the distal pole (Dd). Inset shows an enlarged portion of the dictyosome of the main picture. (From Grove, S. N., Bracker, C. E. and Morré, D. J. (1968) *Science*, 161, 171–3. Copyright 1968 by the American Association for the Advancement of Science. Original print supplied by Professor C. E. Bracker.)

mentioned earlier in this chapter in connection with connective tissue catabolism. It should be remembered, however, that induced $\beta$-galactosidase is thought to be located at intracellular sites (i.e. lysosomes) and lactose, after entering the mycelium, must gain access to the hydrolase before it is further metabolized. One might expect, therefore, that this substrate could possibly be hydrolysed in lysosomes and would imply a heterophagic digestive function for these particles. Further work is needed in this field to clarify the situation.

## 4.4  Extracellular Secretion of Acid Hydrolases by Plants

Most plant organs are able to secrete a wide range of metabolites, but there is little documentary evidence recording the extra-organism secretion of acid hydrolases. Since the observations of Darwin in 1875 it has been known that the leaves of certain insectivorous plants secrete an 'acid ferment' when stimulated by insects or nitrogenous compounds. Although carnivory might be important in the nutrition of such plants, studies on the mechanisms of the digestive process have been neglected and even until very recently a view has prevailed that digestion occurring on the leaf surface may be due to the activities of the saprophytic leaf flora. Recent work has shed new light on the process of digestion in insectivorous plants of the genus *Pinguicula* (Heslop-Harrison and Knox, 1971). These workers have shown that the leaf of *Pinguicula* possesses two types of gland, a stalked gland which is involved in the trapping process through its primary ability to secrete mucilage and the sessile gland which releases, on stimulation, digestive enzymes including acid phosphatase, esterase, ribonuclease and protease. Although both glands contain these enzymes it seems that the secretion of digestive hydrolases is largely confined to the sessile gland which is also involved in the absorption of the products of digestion, presumably via pinocytosis, or a related process. The evidence presented, based primarily on cytochemical procedures at the level of light microscopy, suggests that the head cells of the glands are rich in acid hydrolases which are located largely in the spongy anticlinal walls of these cells although vacuolar sites of location are also indicated in these cells of the sessile glands. It has also been suggested by previous workers that the digestive enzymes in *Dionoea* are stored in the vacuoles of the gland cells, and although all workers in this field to date have avoided use of the term 'lysosome', it is clear that such vacuoles have affinities with these structures although direct biochemical studies have not yet been reported. In addition Heslop-Harrison and Knox used microautoradiographic techniques to follow the fate of absorbed $^{14}$C-labelled protein fed to *Pinguicula* leaves which revealed that digestion products were rapidly absorbed through the sessile glands and accumulated at intracellular sites in the head cells.

One cannot easily avoid making the analogy between this process and the two-way process of extracellular matrix digestion occurring in animal tissues which we discussed above. Clearly the mechanism of secretion and absorption in insectivorous plants would be eminently suited to further examination by ultrastructural and cytochemical methods.

## Suggestions for further reading

CARO, L. G. and PALADE, G. E. (1964) Protein synthesis, storage and discharge in the pancreas exocrine cell, *J. Cell. Biol.*, 20, 473—95.

DINGLE, J. T., BARRETT, A. J. and WESTON, P. D. (1971) Cathepsin D. Characteristics of immunoinhibition and the confirmation of a role in cartilage breakdown, *Biochem. J.*, 123, 1—13.

GROVE, S. N., BRACKER, C. E. and MORRÉ, D. J. (1968) Cytomembrane differentiation in the endoplasmic reticulum—Golgi apparatus—vesicle complex, *Science*, 161, 171—3.

HESLOP-HARRISON, Y. and KNOX, R. B. (1971) A cytochemical study of the leaf-gland enzymes of insectivorous plants of the genus *Pinguicula*, *Planta (Berl.)*, 96, 183—211.

HISLOP, E. C., BARNABY, V. M., SHELLIS, C. and LABORDA, F. (1974) Localization of α-L-arabinofuranosidase and acid phosphatase in mycelium of *Sclerotinia fructigena*, *J. gen. Microbiol.*, 81, 79—99.

HOPKINS, C R. (1969) The fine structural localization of acid phosphatase in the prolactin cell of the teleost pituitary following the stimulation and inhibition of secretory activity, *Tissue and Cell*, 1, 653—71.

HOPKINS, C. R. and BAKER, B. I. (1968) The fine structural localization of acid phosphatase in the prolactin cell of the eel pituitary, *J. Cell Sci.*, 3, 357—64.

KEEN, N. T., LONG, M. and MALCA, I. (1970) Induction and repression of β-galactosidase synthesis by *Verticillium albo-atrum*, *Physiol. Plant.*, 23, 691—6.

PITT, D. (1968) Histochemical demonstration of certain hydrolytic enzymes within cytoplasmic particles of *Botrytis cinerea* Fr., *J. gen. Microbiol.*, 52, 67—75.

SMITH, R. E. and FARQUHAR, M. G. (1966) Lysosome function in the regulation of the secretory process in cells of the anterior pituitary gland, *J. Cell Biol.*, 31, 319—47.

TULKENS, P., TROUET, A. and VAN HOOF, F. (1970) Immunological inhibition of lysosome function, *Nature, Lond.*, 228, 1282—5.

VAES, G. (1969) Lysosomes and cellular physiology of bone resorption, in *Frontiers of Biology*, 14A. *Lysosomes in Biology and Pathology*, vol. 1, ch. 8, pp. 217—53, eds. J. T. Dingle and H. B. Fell. North-Holland Publishing Co., Amsterdam and London.

WOLLMAN, S. H. (1969) Secretion of thyroid hormones, in *Frontiers of Biology*, 14B. *Lysosomes in Biology and Pathology*, vol. 2, ch. 17, pp. 483—512, eds. J. T. Dingle and H. B. Fell. North-Holland Publishing Co., Amsterdam and London.

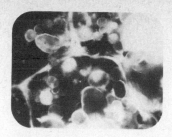

# Chapter 5

# Role of Lysosomes and Acid Hydrolases in Developmental Processes

In this chapter I use the term *development* in the broadest sense to encompass the series of changes occurring in cells, tissues and organs during the life cycle of an organism. *Growth* refers to the *quantitative* changes which accompany development, whilst the *qualitative* changes in cells, tissues and organisms are components of *differentiation*. In some instances such changes have known biochemical bases and these are covered by the term *biochemical differentiation*.

The role of the lysosome system in intracellular digestion in higher organisms is now firmly established following, in particular, extensive work on liver, kidney and polymorphonuclear leucocytes. However, early in the study of lysosomes it became apparent that these organelles were also involved in certain pathological conditions and in processes of natural regression in normal tissues. Spectacular events occur in the postpartum (i.e. the period immediately after birth) involution of the uterus and of the mammary glands when large masses of tissue are rapidly destroyed in a controlled selective manner. Similar dramatic events occur during some embryological processes and in phases of insect and amphibian metamorphosis. All of these changes involve complex tissues which are heterogeneous in respect of cell type and may, in addition, prove difficult to homogenize without disrupting the various organelles. Because of such problems information regarding the role of lysosomes in these processes has been slow in accumulating. As a further consequence of the technical difficulties associated with these tissues it has been impossible to satisfy many of the biochemical criteria relating to lysosomes and much of the evidence rests on cytochemical observations at either the light or ultrastructural levels of microscopy. Furthermore, I should add that developmental sequences and their regulatory mechanisms are complex and comprise an important area of current biological research effort. It would be naive to consider any aspect of development merely in the context of the function of a single organelle. There was a tendency in the early days of lysosome research to explain some developmental phenomena almost exclusively in terms of lysosomal events; my impressions from the current literature suggest that contemporary research workers claim a more modest role of the lysosome system in such processes. The following treatment is of necessity brief, but enough is known to suggest that lysosomes are important in development and warrant a greater research effort.

## 5.1 Involution of the Uterus and Mammary Gland

Although the rapid changes in the uterus and mammary gland during reproduction are massive and involve extensive tissue remodelling the reorganizational processes are under precise control. The coordinated manner whereby large amounts of tissue may be systematically destroyed have attracted the attentions of many workers since the earliest days of lysosome studies. Although such work on both of these tissues, especially the uterus, has been hampered by technical problems of homogenization when relatively low sedimentable hydrolase activity is obtained, substantial information has now accumulated on the role of lysosomes in involution of these organs.

### 5.1.1 In the Uterus
The human menstrual cycle is too well known to require all but the briefest description. This cycle is of twenty-eight days' duration and begins at the commencement of menstruation which normally lasts about five days. Follicle stimulating hormone (FSH), produced by the anterior pituitary gland triggers maturation of the ovum located in the Graafian follicle which is situated within the ovary. At the fourteenth day of the oestrous cycle, under the influence of luteinizing hormone, the ovum is released at *ovulation* and passes to the uterus via the fallopian tube. The empty Graafian follicle becomes the site of development of the corpus luteum which secretes progesterone if fertilization occurs. During these sequential events the glandular mucous membrane lining the uterus (*endometrium*) has thickened following growth in response to oestrogen and progesterone. If fertilization of the ovum occurs it becomes implanted in the endometrium at about the twenty-first day of the cycle and subsequently develops into the embryo. However, if fertilization does not occur the endometrium breaks down and along with the corpus luteum is

released in menstruation. The cycle is basically similar in rodents, upon which most of the experimental observations on the role of lysosomes have been made, except that it is of five days' duration. Combined techniques of biochemistry and ultrastructural histochemistry show only minor changes in total acid hydrolase activity during the cycle but there are marked changes in the *distribution* of lysosomes. Macrophages and other phagocytes become conspicuously active and are rich in lysosomes at the time of endometrial breakdown which prompts the view that these cells play a role in the degradation process. The lysosomes of the epithelial cells are thought to act in an autophagic capacity at this time in rodents, but in humans these tissues appear to be sloughed-off.

The observed coincident changes in lysosomes and their hydrolases with events in the oestrous cycle suggested that the lysosome system may itself be subject to regulation via hormonal control phenomena. Very early work, carried out long before lysosomes were known, demonstrated that several of those acid hydrolases we now know to have lysosome locations decreased in activity about eight days after the removal of ovaries from rats (ovariectomy). A few hours after oestrogen administration to previously ovariectomized animals the levels of lysosomal enzymes in uterine tissues started to increase and within four days had reached a level equal to that occurring prior to ovariectomy. Although this system has not been extensively examined at the ultrastructural level it seems that ovariectomy is accompanied by some tissue involution, perhaps involving autophagy, and during which no new lysosomes are produced so that there is a gradual decline in the population of this organelle, particularly in epithelial cells. Hormone treatments in some way restore lysosomal enzyme synthesis to the normal levels and bring about formation of a new lysosome population. Very interesting results have been published recently (Szego *et al.*, 1971) which throw further light on the effects of steroid hormones on lysosome membranes. Oestradiol, a naturally occurring oestrogen of the vertebrate ovary, was administered intravenously in physiological doses to rats which had been ovariectomized three weeks previously. Fifteen minutes subsequent to hormone injection samples of lysosomes were isolated from the uterus and preputial glands (sex accessory organs rich in lysosomes). When these lysosomes were subjected to autolytic conditions, e.g. detergents and mechanical stress, enhanced lytic susceptibility was found in those lysosomes prepared from these target organs of hormone-pretreated animals. Similarly, administration of diethylstilboestrol labilized the lysosomes of preputial glands of castrated male rats. Also the effect of treatments with testosterone of gonadectomized rats caused destabilization of lysosomes from the preputial glands of either sex. Lysosomes isolated from non-target tissues had not been affected by such steroid injections. The authors contended that the sensitivity, specificity and selectivity of the effects of these steroid hormones on the lysosome membranes suggest that hormone-mediated lysosome labilization may be responsible for the secondary biochemical events occurring in hormone-stimulated target cells and could, therefore, be responsible for a *release* of hydrolases which could then participate in any destruction of target tissues that might occur. This work implies that steroid hormones act at membrane sites rather than primarily at the nucleus as was formerly believed. However, the authors are careful to point out that the possibility of interaction with the nucleus may also occur. In addition, a very recent publication (Badenoch-Jones and Baum, 1973) provides evidence of progesterone-induced permeability changes in rat liver lysosomes *in vitro*. It seems certain, therefore, that this area of research into lysosome surface activation and response to reproductive hormones is due to yield important results in the near future.

There are a number of reports suggesting increased total amounts of lysosomal enzymes in the uterus during its extensive phase of growth in pregnancy but no firm conclusions have yet been drawn from the observations made at early stages of pregnancy. Towards the end of pregnancy a number of possible roles of lysosomes are envisaged in which phagocytosis may be involved, e.g. degradation of tissues during formation of the birth canal or late pregnancy effects on the endometrium which becomes detached with the placenta, but little firm information relating lysosomes to such processes is available at present. The most striking example of tissue destruction in the uterus occurs during the period immediately following birth. At this time of postpartum involution many complex tissues and macromolecules are broken down at an enormous rate in a precisely controlled fashion. Most of the available information has been derived from studies on rats in which 90 per cent of the collagen of the uterus is destroyed within 96 hr postpartum. In this organism the activity of lysosome hydrolases within uterine tissues increases several-fold and reaches a peak at four days' postpartum when the process of involution is nearing completion. These apparent increases in hydrolase activity are due to the level of tissue hydrolases remaining constant as tissue fresh weight decreases. Accompanying the progressive relative increase in hydrolase activity is the arrival of macrophages, derived both from tissue histiocytes and monocytes of the blood, which show intense lysosomal hydrolase activity. These macrophages are distributed throughout the various tissues of the uterus and participate in vigorous phagocytosis of collagen which may have been previously partially degraded by extracellular collagenase. However, there is no substantial evidence indicating labilization of lysosomes leading to autolytic cell death; it would appear that this organelle is, however, very active in intracellular digestive processes involving autophagy. The involution process is also accompanied by a great decrease in the main component of the uterus, smooth muscle, which appears to be achieved by a process involving the reduction in cell volume rather than of cell number, but the mechanism of this facet of involution is obscure.

## 5.1.2 In the Mammary Gland

Lysosomes are found in the resting mammary gland where they may be readily detected by cytological methods, but with much more difficulty by biochemical means. Their exact function in resting glands is unknown but one must suspect, in the absence of evidence to the contrary, that they serve the same role as that established in other tissues, namely, intracellular digestion and turnover of macromolecules. At the onset of pregnancy the concentration of lysosomal enzymes measured on a fresh weight basis increases even as the weight of the enlarging gland increases. The necessity for such increased enzyme concentration is not understood at present, but it may be associated with the remodelling necessary as the organ increases in size, and with cell differentiation which heralds the process of lactation. It would appear that all cell types of the gland contain lysosomes but the secretory cells are especially rich in lysosomal hydrolases, particularly acid phosphatase. Indeed lysosomal enzymes are found in secretions from the mammary gland although any reason for this is not yet apparent.

The changes accompanying mammary involution are rapid and spectacular and involve dramatic remodelling of tissues. During the process, which may be conveniently induced by removal of the suckling offspring, there is an apparent increase in the acid hydrolase content of tissues which is largely due to the reduction of tissue mass which is unaccompanied by a corresponding decline in activity of these enzymes. Earlier findings by Greenbaum and Slater (1965) appeared to suggest that there was a release of particle-bound hydrolases during the involution process, but the current belief is that this was an artefact due rather to the increased fragility of the lysosomes which were thereby predisposed to damage during tissue fractionation. Much of our recent knowledge has been obtained from electron microscope studies rather than from biochemical observations which are difficult owing to the problems encountered with tissue fractionation. On the first day of induced involution in rodents large vacuoles are present in the gland and contain lamellated structures. By the third and fourth days of involution the number of vacuoles decrease with a gradual increase in myelin figures, and by the tenth day residual bodies predominate. The conclusions from this careful work were that lysosomal enzymes were not released in quantity but rather that lysosomes retained their integrity and participated in extensive autophagic activity, particularly in the epithelial cells. Macrophage activity increases markedly in the process of involution and is responsible for the removal of the detritus of dead cells and milk. It is also apparent, however, that a number of products of cell lysis are removed from the sites of involution via the capillaries and the lysosomes are not the exclusive agents of tissue resorption in the mammary gland. It is also recognized that the mammary involution process, like that of the uterus, is under the subtle control of hormonal regulatory mechanisms about which we are largely ignorant.

## 5.2 Lysosomes in Metamorphosis

Of all the processes of development that are subject to hormone regulation few are as impressive as those changes accompanying metamorphosis. During metamorphosis in insects and amphibians massive amounts of larval tissues are destroyed in a highly organized and sequential manner and the molecular components are re-utilized as a major source of components in the *de novo* synthesis of materials comprising the tissues characteristic of the adult form. It is known that metamorphic phenomena are under the influence of external environmental factors including daylight, daylength and temperature. The appropriate environmental signals are transmitted through the nervous system to either the hypothalamus or neurosecretory cells. Such stimuli in turn influence the production of 'brain' hormones which act on the pituitary in the case of tadpoles and on the corpora allata and prothoracic gland in insects. In the case of amphibians, thyrotrophic hormone from the pituitary gland activates the thyroid gland which produces the effective metamorphic hormones, thyroxine and tri-iodo-thyronine, whilst in insects ecdysone is the hormone ultimately responsible for triggering the metamorphosis. However, these hormones alone are not adequate to control metamorphosis and a delicate balance exists between them and certain other endocrine secretions. In insects such a balance existing between 'juvenile hormone' and ecdysone regulates the manner and progress of metamorphosis. It is believed that a prolactin produced by the tadpole pituitary may also function in conjunction with thyroid stimulating hormone (TSH) and TRH (TSH releasing hormone of the hypothalamus) in the feedback regulation of hormone-induced metamorphosis in amphibians. The metabolic bases of metamorphosis in both these groups of animals is undergoing intensive investigation at the present time and fortunately experimentation is facilitated by the ease with which the processes may be triggered by the appropriate use of hormones. Amphibians are particularly suitable for studies on regression since the tail tissues may be maintained in organ culture and induced to undergo involution *in vitro* in the presence of thyroid hormones. The biochemical effects of applying hormones are dramatic and apart from the obvious external signs of morphological change such as loss of the tail and emergence of the limb buds in amphibians and the formation of wings in insects there are profound internal changes within organs to fulfil the needs of the new modes of life and changing diets. For those students wishing to acquaint themselves more fully with the morphological and biochemical changes of metamorphosis in these two groups Tata (1973) provides a stimulating general essay on the topic and a more extensive coverage (Tata, 1971) of hormonal regulation in metamorphosis.

Of the many changes which sweep through the tissues of organisms undergoing metamorphosis those involving tissue regression have received considerable attention following establishment of the lysosome concept. It

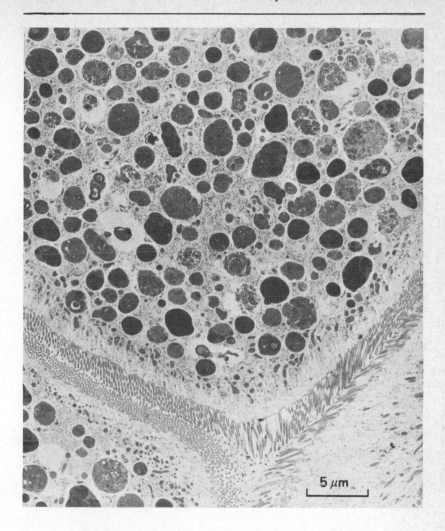

**Fig. 5.1**    Effect of synthetic moulting hormone, ecdysterone, on midgut cells of the flesh-fly, *Sarcophaga bullata*. Electron micrograph of a thin section of an apical portion of a midgut epithelial cell 9 hr after ecdysterone injection of the larva. The hormone causes a massive increase in the size and number of secondary lysosomes and accelerates the effect observed at the onset of natural metamorphosis which would normally occur 48 hr later and would show extensive cellular autophagy similar to that induced by the hormone. Control larvae injected with saline show a low number of lysosomes which are inconspicuous. (From Radford, S. V. and Misch, D. W. (1971) *J. Cell Biol.*, 49, 702–11. By courtesy of The Rockefeller University Press. Original print supplied by Professor D. W. Misch.)

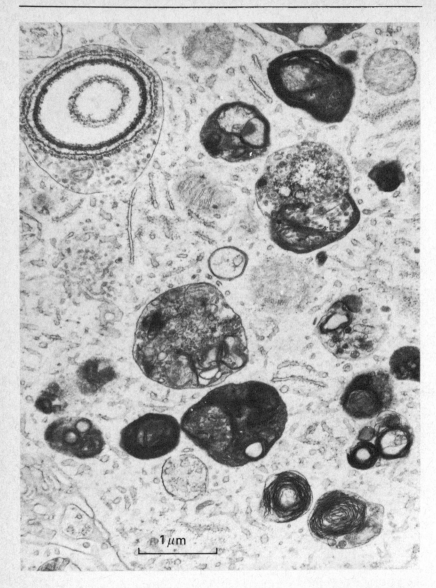

**Fig. 5.2** Electron micrograph of the subject of Fig. 5.1, larval cells of *Sarcophaga bullata*, at a higher magnification and showing details of the ecdysterone-induced secondary lysosomes. The variety of sub-cellular components identifiable in these structures indicates the extensive nature of cellular autophagy in induced metamorphic phenomena. Images seen in larvae undergoing normal metamorphosis are identical. (Original unpublished print supplied by Professor D. W. Misch.)

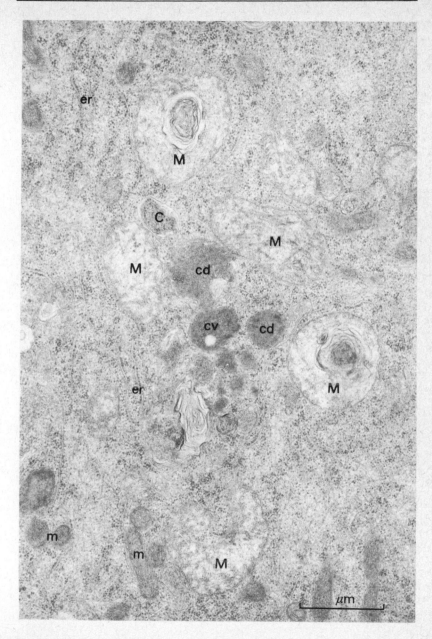

**Fig. 5.3**  Role of lysosomes in metamorphosis. Localization of lytic activity in the prothoracic gland of the oak silkworm, *Antheraea pernyi* during the prenymphal

stage. Autophagic activity is pronounced and occurs in secondary lysosomes (M) which show myelin figures and whorls of membranous materials. There is a marked increase in the frequency of autophagic bodies during metamorphosis. Vacuolated bodies (cv) and dense bodies (cd) also occur and were shown to contain esterase and phosphatase activity by application of the Gomori procedure at the ultrastructural level. These bodies are considered as equivalent to lysosomes and to possess lytic capabilities. m, mitochondria. (From Beaulaton, J. (1967) *J. Microscopie*, 6, 179—200. By courtesy of La Société Française de Microscopie Electronique. Original print supplied by Dr J. Beaulaton.)

---

has been known since the early days of studies on lysosomes that the processes of resorption of the tadpole tail and regression of the various insect larval tissues, including the fat body and the salivary gland, are accompanied by increased lysosomal acid hydrolase activity. However, research is still hampered by the failure to extract significant quantities of particle-bound hydrolases from these tissues and to separate the specific cell types for biochemical examination. The more recent use of electron microscope cytochemistry on these subjects has enabled certain important conclusions to be drawn regarding the processes, and it is now clear that widespread solubilization of lysosomal enzymes does *not* occur during tissue resorption but in most cases where the events have been critically studied in insects it would seem that enhanced autophagy involving acid hydrolases plays a prominent role in the sequestration and digestion of cytoplasmic components (Figs. 5.1, 5.2, 5.3). From the pioneering studies on the role of lysosomes in amphibian metamorphosis, carried out in the early days of lysosome work, there persists the lingering belief that much of the increased acid hydrolase activity accompanying the process arises by activation of preformed lysosome enzymes. More recent work, utilizing inhibitors of protein synthesis, has tended to cast some doubt on these assumptions. Such work showed that various inhibitors, including actinomycin D, cycloheximide and puromycin, irreversibly prevented thyroid hormone induced involution of tadpole tails *in vitro*. Also suppression of tail regression by actinomycin D and cycloheximide prevented the incorporation of tritiated uridine into RNA, and puromycin and cycloheximide abolished the uptake of $^{14}$C-amino acids into new proteins (Fig. 5.4). However, the current climate of biological opinion regards with considerable caution the results obtained using inhibitors of protein synthesis because these compounds are known to have widespread metabolic side effects which may be little related to protein synthesis. Perhaps particularly apposite examples in the present context are seen in the multiple effects of puromycin (thought to cause abortive protein synthesis by acting as an analogue of the terminal group of phenylalanyl *t*-RNA) on tadpoles which included *reduction of the activity* of phospho-diesterase and *activation* of glycogen synthetase (Blatt, *et al.*, 1969). A more superficial and incorrect interpretation of these observations could

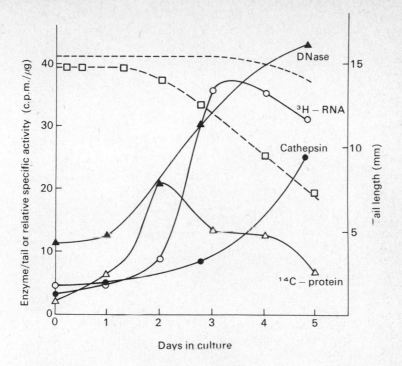

**Fig. 5.4** Changes in activity of hydrolases during regression of tadpole tails. Increased activity of cathepsin (●) and deoxyribonuclease (▲) and simultaneous increase of RNA and protein synthesis during induced regression of tadpole tails in organ culture caused by the addition of tri-iodothyronine to the medium. ○———○, incorporation of [³H] uridine into RNA; △———△, incorporation of ¹⁴C-labelled amino acids into protein. Incorporation on day 0 is the average value for controls; all other points refer to samples to which 3,3¹,5-tri-iodo-L-thyronine was added. ———, tail length of controls over the duration of the experiment; □———□, length of tri-iodothyronine-treated tails showing a marked onset of regression between the second and third day of culture. These observations have been taken as presumptive evidence for the *de novo* synthesis of cathepsin and deoxyribonuclease, but no evidence is yet available to show a simultaneous increase in the amount of respective enzyme proteins or specific incorporation of label into such proteins. (From Tata, J. R. (1971) *Symp. Soc. exp. Biol.*, **25**, 163–81. By courtesy of the Society for Experimental Biology.)

invoke explanations based on presumed synthesis of enzyme protein *de novo*. Notwithstanding these limitations associated with the use of inhibitors of protein synthesis the weight of evidence is now beginning to indicate that programmed cell death occurring in regression is dependent upon the synthesis of some new proteins and that hormones such as

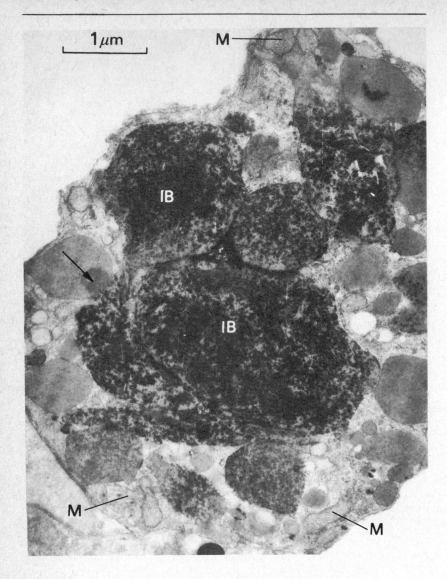

**Fig. 5.5** Ultrastructural events in regressing tail muscles of *Xenopus* larvae during metamorphosis. Well developed macrophage from a tail rudiment stained for acid phosphatase activity. Inclusion bodies (IB) differ in size and enzyme content; those with a positive reaction for acid phosphatase may be regarded as 'phagosomes'. Some inclusion bodies contain myelin figures (arrow), i.e. secondary lysosomes. Small mitochondria (M).

Early stages of tail regression involving massive reorganization of muscle fibres may be independent of lysosome participation but later stages appear to involve extensive scavenging by macrophages rich in lysosomes which contain high hydrolase content. (From Weber, R. (1964) *J. Cell Biol.*, **22**, 481—7. By courtesy of The Rockefeller University Press. Original print supplied by Professor R. Weber.)

ecdysone seem to activate dormant genes of insect salivary glands. Other work on the hormone-induced resorption of tadpole tails shows that part of the complement of increased activity of cathepsin and deoxyribonuclease which arises is due to newly formed enzymes which possess properties different from those of pre-existing lysosome enzymes. It is currently believed that much of this synthesis of lysosomal enzymes in regressing tail tissues is associated with the enhanced macrophage activity (Fig. 5.5), which characterizes this phase of the process of metamorphosis.

## 5.3 Lysosomes during Sponge Development

The sponges, which constitute the phylum Porifera, are the most primitive of multicellular animals. Sponges vary greatly in size and form but essentially they are composed of relatively loose aggregations of several cell types comprising the mesenchyme which is covered by a surface epithelium. The epithelial layer is perforated by ostia that serve for the entry of water which after being driven through the interior of the animal in a water-stream created by action of flagella of the choanocytes lining the interior, passes to the outside via other orifices. A skeletal system is present in the mesenchyme and is composed of needle-shaped spicules of silicic acid.

Only the fresh water sponges, particularly *Ephydatia mulleri*, have received any detailed attention with regard to lysosomes, and the remainder of this section, therefore, refers to this or related species. The mesenchyme consists of several amoeboid cell types, or amoebocytes, embedded in a gelatinous matrix which is supported by skeletal components. Several types of amoebocytes are recognized and of these the archaeocytes, cells rich in food reserves, and capable of differentiation into other types of cell, i.e. they are totipotent, have received most attention. A further cell type, the granular amoebocyte or cystencyte has also been the subject of some lysosome studies. During autumn the European freshwater sponges produce large numbers of gemmules each consisting of archaeocytes rich in food reserves and surrounded by other amoeboid cells which are encased in a resistant shell impregnated with spicules. The gemmules are released on disintegration of the parent sponge and serve as resistant propagules. Following a maturation period the gemmules germinate in the spring when the cells of the interior emerge through a micropyle and differentiate into the various cell types characteristic of the adult sponge.

All the cell types so far examined contain acid hydrolases localized in lysosome-like structures, but since these enzymes in sponges do not withstand the rigours of the fixation procedures of electron microscopy, ultrastructural observations dependent on cytochemical procedures have been limited and until the present largely confined to the germinating gemmules and the cystencytes. Only limited biochemical observations have been made to date. Under experimental conditions the gemmules may be induced to germinate after storage at 3° C by transfer to water at 20° C for a few days. During the ensuing early phases of differentiation no new food material is taken in and the stored proteins, ribonucleoproteins and lipids of the large lens-shaped cytoplasmic inclusions, known as vitelline platelets, are mobilized and used in biosynthetic processes. Digestion of the vitelline platelets has been extensively studied by light and E.M. cytochemistry in the predominant cell type, the archaeocytes, which emerge from the gemmules. Such observations reveal the appearance within the cytoplasm of lysosomes which transfer lytic enzymes to the vitelline platelets which subsequently become degraded. Cystencytes arise somewhat later in the differentiation process and the vitelline platelets within these cells are digested following their inclusion within large hydrolase-containing autophagic vacuoles.

Other stored products such as lipids and glycogen occur within inclusions in the cytoplasm of most cell types of sponges, but at present there is no strong evidence for the involvement of lysosomes during digestive events associated with these structures.

## 5.4 Lysosomes and Acid Hydrolases in Plant Development

In the earlier part of this chapter I examined the role of lysosomes and acid hydrolases in the controlled selective destruction of tissues in some developmental processes in animals. Increasingly frequent reports are now appearing in the literature relating changes in the locations and activities of acid hydrolases to developmental phenomena in plants. Progress in this field of endeavour has been hampered by the shortage of information on the existence and nature of lysosomes in plants. Now that methodological problems are being overcome the way has been opened for more detailed assessments of the role of lysosomes in plant development. However, it should be borne in mind that acid hydrolases in plants, probably even more so than in animals, are also found at extralysosomal sites, e.g. cell wall, nucleus, plasmodesmata, and involvement of acid hydrolases may not necessarily imply participation of the lysosome system in these processes.

### 5.4.1 In Roots

The earliest observations were made by Gahan and Maple (1966) who demonstrated by histochemical means the occurrence of lysosome-like particles in the undifferentiated meristematic cells of the root tips of broad bean (*Vicia faba*). By use of the Gomori technique for β-glycerophosphatase they were able to show by light microscopy that the onset of cell differentiation leading to the formation of protoxylem vessels was accompanied by a change in the location of this hydrolase from particulate sites to the cytoplasm. The manner in which these changes are related to the complex series of events involved in such differentiation, even though in this case it terminates in the production of dead xylem elements, is obscure at present. Matile (1969) considers that these particular observations reveal *formation* of the lysosome apparatus rather than its disruption; a view which could only be clarified by careful observations at the ultrastructural level using cytochemical techniques for hydrolase localization. Similar observations to those of Gahan's group have been made during the cataclysmic death of root cap cells when the lysosomal location of acid phosphatase gave way to a diffuse cytoplasmic location at the onset of autolysis. In addition these observations have been confirmed in the root cap of *Zea mays* by cytochemical techniques for acid phosphatase localization at the level of ultrastructure (Berjak, 1968) which also showed that widespread disorganization in the compartmentation of the cell accompanied the process. Although the histochemical evidence is strong in these two examples of root differentiation there are negligible biochemical data to support these findings at present. Unfortunately this unhappy state of affairs is likely to persist for some time since root tissues, in common with most plant organs, are complex structures that are heterogeneous with respect to cell types which cannot be separated from each other in quantities suitable for biochemical analysis.

For many years plant anatomists have maintained that emergence of endogenously produced lateral roots arising in the pericycle occurs through mechanical rupture of the cortical tissue. Some plant physiologists hold the additional view that the emergence of the roots is facilitated by the action of hydrolytic enzymes upon the cortical tissues in the immediate vicinity. Support for the latter view has been provided by a rather elegant experiment by Sutcliffe and Sexton (1968) who showed increased β-glycerophosphatase activity in those serial basipetal sections of pea roots which contained lateral root primordia. Furthermore, such increased enzyme activity could be demonstrated histochemically at specific cortical sites adjacent to the lateral root primordia in the pericycle. Although the original authors did not invoke a lysosomal explanation of the process it is possible that lysosomes may be involved and it would seem that this is an area worthy of ultrastructural studies.

## 5.4.2   In Stems
Little information is available to date which relates lysosomes to developmental processes in stems. Much of the work in our laboratory is focused on the changes in lysosomes and lysosomal enzymes during mechanical

damage, stress and fungus infections of potato tubers and we shall encounter this again in Chapter 6. With the recent development of a technique for isolating potato sprout lysosomes in experimental quantities (Pitt and Galpin, 1973) it should now be possible to extend the method to other materials, e.g. seedling stems which prove such useful material for many studies in developmental plant physiology. This is an area of research which could prove fruitful for those seeking to relate lysosome function to developmental processes.

### 5.4.3 In Leaves
Although the early findings of Walek-Czernecka were made using onion bulb scale leaves, relatively few observations have been made on the occurrence of lysosomes in green leaves. The sparse literature contains electron micrographs of sections of both monocotyledonous and dicotyledonous leaves showing autophagic bodies and lysosome-like objects, but there are few reports of work on isolation and characterization of such particles. However, the author (Pitt, 1973) has succeeded in isolating and characterizing a lysosome fraction from potato leaves and has examined the behaviour of this fraction during disease caused by the potato late blight fungus, *Phytophthora infestans* (see Chapter 6, p. 139). On the other hand, however, considerable information has accumulated about the behaviour of some acid hydrolases, generally considered to have lysosomal locations, during leaf damage and senescence. Both of these processes are accompanied by increased activity of acid hydrolases with huge increases in ribonuclease activity in leaves of several species (see Pitt and Galpin, 1971, for a review). Current observations in our laboratory suggest that at least a portion of the increased ribonuclease activity arising during mechanical damage to potato leaves may be associated with the lysosome fraction. Whether this involves activation of pre-formed enzyme molecules or *de novo* enzyme synthesis is a problem of current research interest in several laboratories. It would also be valuable to learn if the processes leading to enhanced acid hydrolase activity during stress and senescence have fundamental similarities to those changes in the properties of lysosomes already discussed in relation to aging and starvation in *Euglena* and *Tetrahymena* in Chapter 3.

### 5.4.4 In Flowers
The existence of lysosomes in pollen grains and pollen tubes has already been mentioned previously (Chapter 2, p. 39), and was reported in early work which helped to establish the existence of lysosomes in plant tissues. It should also be emphasized that acid hydrolases are by no means confined to the vacuole system in pollen and occur in appreciable quantities in the inner wall or intine of the grains. Few facts are available relating the acid hydrolases in pollen tissues to possible functions, but it may be that they are involved in mechanisms whereby pollen tubes penetrate the tissues of the ovary during fertilization.

Acid hydrolases have also been located in seed embryos, but precise information concerning these hydrolases in other flower parts is not currently available except for the interesting observations on corolla segments made in Matile's laboratory in Switzerland. It has been known for many years that wilting of petals is associated with a rapid fall in their protein content: an observation providing presumptive evidence of lytic processes in the cells. Recent work in plant science has conclusively demonstrated that processes of senescence involve not only breakdown of protein but also a simultaneous protein synthesis, the balance between which regulates the progress of senescence. Such observations strengthened the earlier views of Gahan, Matile and others that lysosomes exist in plant cells and serve to segregate lytic compartments from those involved in biosynthetic processes.

The corollas of the various species of bindweed provide a familiar and striking example of rapid senescence in floral parts. Mature flowers of the Morning Glory (*Ipomoea purpurea*) open in the early morning, expand rapidly about noon, start to wilt in the afternoon and are completely senesced by the next morning. Matile and Winkenbach (1971) followed by means of biochemical and ultrastructural techniques the changes in the activity of acid hydrolases and the breakdown of cytoplasmic organization in the wilting corolla of this species. During the wilting period there is a rapid disappearance of important macromolecules from the tissues including nucleic acids and proteins which can be partially correlated with the rapid increases in some acid hydrolases, particularly DNase (Fig. 5.6). Although increases in activity of certain hydrolases are prevented by the presence of inhibitors of protein synthesis these workers are careful to point out that the protein synthesis necessary for the process may not be in respect of specific hydrolase proteins. The ultrastructural evidence provided is fairly indicative that the large vacuole of the mesophyll cells represents a lysosome compartment. Invagination of the vacuolar membrane (tonoplast) is accompanied by enclosure of cytoplasmic materials which are presumably hydrolysed by acid hydrolases (Fig. 5.7). Subsequent autolysis is then believed to follow a highly organized pattern during which the hydrolysis products leave the floral area via the phloem and vascular elements which remain functional until the majority of breakdown products have been transported to sites of developing flower buds and seeds. In this case then the central vacuole of the mesophyll cell represents a lysosome which can become multivesiculate during the autophagic process. It is now believed by some workers that membrane-enclosed materials that initially enter plant cells by pinocytosis may likewise become incorporated into the central vacuole during heterophagy (Mahlberg, 1972a, b). So far, however, it has not been possible to isolate membrane-associated acid hydrolases from corollas of Morning Glory nor is there any histochemical evidence pertaining to the ultrastructural location of these hydrolases. Consequently, conclusive proof of the

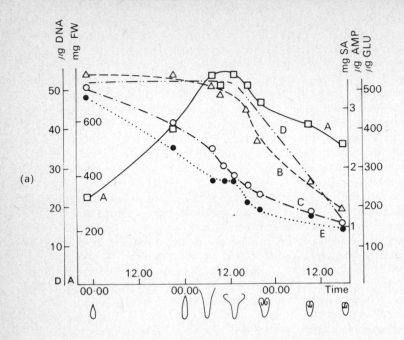

(a)

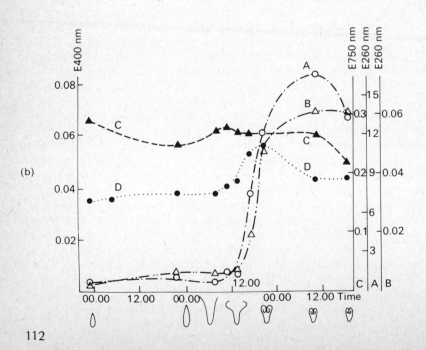

(b)

◀ Fig. 5.6    Changes in the content of some macromolecules and hydrolases of corollas of the Morning Glory (*Ipomoea purpurea*) during senescence.

(*a*)    Content    in    (A □ ———— □)    freshweight;    (B △ ————— △)    protein; (C ○ — · — · — ○)    RNA;    (E ● · · · · · · ●)    starch;    and    (D — · · —)    DNA    per corolla.

(*b*)    Relative    activities    of    acid    hydrolases    per    corolla:    (A ○ — · — · — ○)    RNase; (B △ — · · — △)    DNase;    (C ▲ ————— ▲)    protease,    and    (D ● · · · · · ●) α-glucosidase.

The sketches below the time axis indicate the shape of the corolla at these times.

The striking fact emerges that the decrease in DNA content of senescing corollas coincides with increasing DNase activity. It is not yet clear if this increased enzyme activity is due to the *de novo* synthesis of enzyme protein or to activation of pre-formed enzyme. (From Matile, P. and Winkenbach, F. (1971) *J. exp. Bot.*, 22, 759–77. By permission of Oxford University Press.)

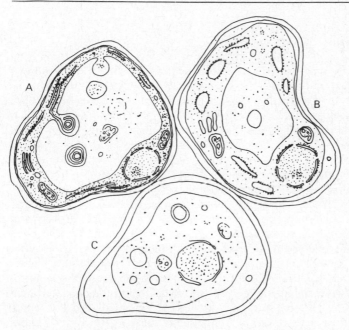

Fig. 5.7    Schematic diagrams of the changes in the ultrastructure of mesophyll cells during senescence of the corolla of *Ipomoea purpurea* (Morning Glory).

(*a*)    Autophagic activity of the vacuole; invaginations of the tonoplast result in sequestration of cytoplasmic material into the lysosomal compartment (i.e. the central vacuole according to the interpretation of Matile).

(*b*)    Shrinkage of the vacuole, dilution of the cytoplasm and inflation of cytoplasmic membrane systems.

(*c*)    Autolysis is initiated by ultimate breakdown of the tonoplast.

The concept of the central vacuole of plant cells as a large lysosome is an attractive notion held by Matile which has not yet been examined in detail by other workers. (From Matile, P. and Winkenbach, F. (1971) *J. exp. Bot.*, 22, 759–77, 1971. By permission of Oxford University Press.)

involvement of lysosomes in the wilting of flower petals awaits further experimentation.

### 5.4.5 In Fruits

Although an attractive hypothesis has persisted in botanical circles for some time that ripening of fruits may involve the lysosome system, the only publication on this topic describes the biochemical characterization of lysosomes from tomato fruits (Heftmann, 1971). Little information is available concerning any role that the lysosome system may play in the tomato fruit, but there is evidence that the sedimentability of lysosomal esterase increases in the ripening process. There are no indications so far that a release of lysosomal enzymes occurs during the process.

### 5.4.6 In Seeds

Seeds usually contain stored materials which become mobilized during germination through enzyme-mediated hydrolysis, with the breakdown products being translocated to and used by the developing embryo. In dicotyledons food is usually stored in the cells of the cotyledons as lipid droplets, starch grains and in protein-rich aleurone grains (Fig. 5.8) which also contain some acid hydrolases. In monocotyledons reserves are generally stored in the endosperm and the aleurone grains. Frequently a particular seed type has a single storage tissue, but more than one type may occur in some seeds, e.g. some dicotyledons contain endosperm in addition to the reserves in the cotyledons.

Perhaps the most extensively examined germination system is that involving the relationships between endosperm hydrolysis, hydrolases and growth hormones in cereal seeds. If the embryo is removed from barley seeds then the endosperm reserves are not mobilized. It is apparent that the embryo controls the hydrolysis process by means of which the reserves are made available for subsequent translocation to the developing embryo. The hydrolysis of starch is due mainly to the activity of α-amylase which occurs only in low concentrations in the ungerminated seed but which appears in high activity in the endosperm during germination. Other enzymes behaving similarly include proteases, ribonucleases and 1,3-β-glucanase which are all involved in the breakdown of minority substrates of the endosperm. Over ten years ago reports appeared independently (Paleg, 1960; Yomo, 1960) of the effects of gibberellic acid ($GA_3$), a plant growth hormone, on cereal endosperm digestion in embryo-less seeds. Removal of the embryo from untreated seeds prevents the process of endosperm digestion since the embryo is the major source of endogenous gibberellic acid which in conjunction with other growth hormones triggers and regulates the hydrolysis. It soon became apparent that isolated aleurone layers had the capacity to respond to added $GA_3$ by producing and liberating large amounts of the various hydrolases to the incubation

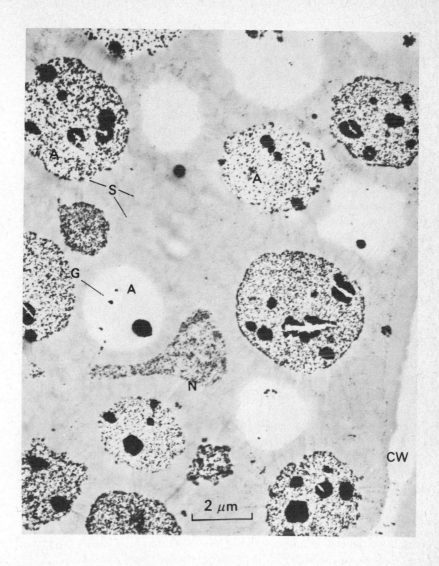

**Fig. 5.8** Electron micrograph showing localization of acid phosphatase in the cells of cotton seed cotyledons. Deposition of lead phosphate occurs faintly throughout the ground substance, conspicuously in the nucleus (N) and aleurone grains (A) and is absent from the spherosomes (S) and certain aleurone grains. CW = cell wall;

G = globoid. (From Yatsu, L. Y. and Jacks, T. J. (1968) *Archiv. Biochem. Biophys.*, 124, 466—71. By permission of Academic Press Inc. Original print supplied by Dr L. Y. Yatsu.)

medium thus establishing that the aleurone cells were the sites of enzyme production. The process as envisaged in germinating barley seeds is represented schematically in Fig. 5.9.

Early views favoured the idea that activation of preformed hydrolytic enzymes occurred during the process of endosperm digestion, but classic work on enzyme protein synthesis in this system by Filner and Varner (1967) involving experiments with $^{14}$C-amino acid incorporation and density-labelling with heavy water demonstrated conclusively that *de novo* synthesis of α-amylase occurs in the germination process. Recent work, based largely on studies with inhibitors of protein synthesis, indicated that many of the other hydrolases occurring at increased levels during germination may also be newly synthesized and follow the pattern established for α-amylase production which involves gibberellin-mediated de-repression of appropriate genes in the aleurone cells. Superficially, acid phosphatase shows a similar response in $GA_3$-treated half-seeds (i.e. embryo-less seeds) and aleurone tissue to that for α-amylase (Fig. 5.10), but in the case of the former enzyme high levels of activity are found in ungerminated seeds and are associated with vesicular sites in the aleurone grains. It may be that in the case of acid phosphatase alternative mechanisms are involved in increasing enzyme activity during germination. Although *de novo* synthesis of the *secreted* enzymes is thought to be responsible for much of the digestion of the endosperm in cereals, it is also necessary to provide an explanation for the digestion of the reserve materials occurring in the aleurone grains themselves and which are rich in stored phosphate in the form of phytin (calcium inositol hexaphosphate) and proteins. One must presume that the hydrolases found in the aleurone grains, and considered by some workers as located therein in particles equivalent to lysosomes, digest the reserves in the aleurone grains and supply those materials required in the *initial* reorganization of the aleurone cell, involving ribosome and membrane biosynthesis, which provides the machinery participating in synthesis of new enzyme protein that may then be secreted to and participate in the digestion of the endosperm. Critical examination of aleurone cells by E.M. cytochemistry at early stages in the germination process reveals that those vesicles of the aleurone cells rich in acid hydrolases may fuse with the cell membrane on the side of preferential enzyme secretion adjacent to the endosperm. This phenomenon is very reminiscent of the process of extracellular secretion of lysosome enzymes mentioned in Chapter 4, p. 85. It seems that this secretory process would not primarily involve the extrusion of the pre-formed or newly activated lysosomal enzymes initially present in the aleurone grains, but is concerned with the transport of the newly synthesized hydrolase component to the endo-

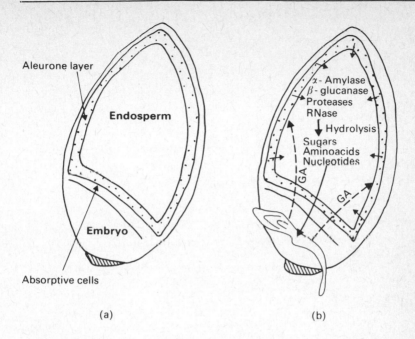

(a)  (b)

**Fig. 5.9** Sequence of events envisaged in the mobilization of stored reserves of the endosperm of barley seed during germination. Gibberellins pass from the embryo and induce *de novo* synthesis of important hydrolytic enzymes, e.g. α-amylase, glucanase, protease and RNase in the aleurone cells. These act upon the macromolecular components of the endosperm giving rise to breakdown products which are translocated to the germinating embryo. It is possible that some hydrolases are already present in the dormant seed either in an active or inactive state and these may also play a part in the digestion of the endosperm in addition to mobilizing the reserves located in the aleurone grains. (*a*) Ungerminated seed; (*b*) germinating seed.

sperm. However, the fairly high activity of acid phosphatase in dry seeds may indicate that the secretory process could convey some pre-formed phosphatase to the endosperm.

Not everyone considers that lysosomes are involved in the process because few biochemical data have yet become available in support of the ideas of Gibson and Paleg (1972) who produced evidence on the lysosomal location of GA-inducible enzymes in wheat aleurone cells. Convincing proof for or against this hypothesis awaits the isolation and characterization of hydrolases from both the aleurone grains and those vesicles which may contain newly packaged and activated enzymes. However, fairly good biochemical evidence is now available confirming the ultrastructural findings of lysosome-like structures rich in acid phosphatase and other hydrolases in the cotyledons of germinating pea, tobacco and cotton seeds,

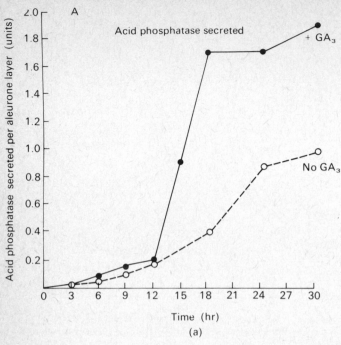

(a)

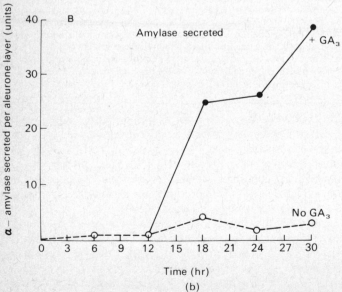

(b)

◄ **Fig. 5.10**  Secretion of acid phosphatase and α-amylase by detached barley aleurone layers.

(a)  Acid phosphatase occurs in the aleurone cells of ungerminated barley seeds. Detached aleurone layers secrete some acid phosphatase in the absence of added gibberellic acid ($GA_3$), but in the presence of $GA_3$ the secretion is greatly enhanced. In the aleurone layers acid phosphatase is located in the aleurone grains within particles having affinities with the lysosomes. It may be that pre-existing enzyme is secreted following activation or *de novo* enzyme synthesis. It is also possible that both processes occur. In germinating intact seeds this enzyme would be secreted to the endosperm and presumably hydrolyse reserve phosphate-containing substrates. Initially the enzyme may also be involved in mobilizing aleurone reserves.

(b)  The secretion of α-amylase from barley aleurone layers. The ungerminated seed contains negligible α-amylase and that which is secreted arises *de novo*. It is not clear at present if this enzyme is located at and secreted via particulate sites. Some support exists for the idea that newly synthesized enzyme is packaged and secreted in membrane-bounded bodies (lysosomes). (Data provided by Miss Kathryn Pickup, University of Exeter.)

---

and it is conceivable that the digestion of reserves in these tissues and those of the barley endosperm may be linked to intracellular lysosome activity.

## 5.5  Lysosomes in Developmental Processes in Protists, Particularly Fungi

I have already mentioned changes that occur in the activity and locations of lysosomal hydrolases during the growth cycle of *Tetrahymena* (Chapter 3, p. 70) and now I wish to consider some work which relates to development in fungi. Microorganisms have proved valuable subjects for studies on biochemical differentiation since they can often be grown under precisely controlled conditions in continuous or synchronous culture. These morphological and biochemical changes associated with the growth cycles of microorganisms involve turnover of cellular components. Macromolecules, including proteins, which are no longer required are re-cycled by the cell and the products of their intracellular digestion are used in bio-synthetic processes to fulfil the specific macromolecular requirements of developmental change. There is now some evidence that biochemical differentiation in some eucaryotic protists, e.g. yeasts, may be correlated with the activities of certain lysosomal enzymes.

### 5.5.1  Role of Lysosomes and Hydrolases during Development in Yeasts

Diauxic growth is a phenomenon displayed by some microorganisms grown in pure culture in the presence of two different carbon sources. This situation is exemplified by the classic case of *Escherichia coli* grown on a medium containing carbon as a mixture of glucose and xylose. Glucose is

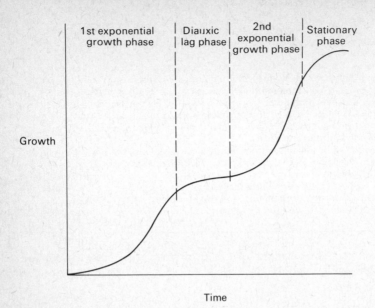

**Fig. 5.11** An outline of the main features of the diauxic growth curve in micro-organisms as exemplified by utilization of a dual carbon source. During the first exponential growth phase the first carbon source is preferentially utilized. In the diauxic lag phase it is thought that redundant enzymes associated with the metabolism of the first carbon source are re-cycled and inducible enzymes required for utilization of the alternative carbon source are synthesized. A second exponential growth phase occurs until exhaustion of the second carbon source. A stationary phase occurs when both carbon sources are exhausted.

used preferentially and is rapidly exhausted. After this point is attained there follows a lag phase when growth falls off. During this transient lag phase (the diauxic lag phase) inducible enzymes are synthesized permitting utilization of the alternative carbon source which is then reflected in an upsurge in growth rate (the second exponential growth phase) that ceases when the second carbon source is exhausted (Fig. 5.11). When particular strains of *Saccharomyces cerevisiae* are batch-cultured on glucose-containing medium diauxic growth occurs during which glucose is initially metabolized and exhausted fermentatively to give carbon dioxide and ethanol. After a while a diauxic lag phase features and growth is resumed at the expense of the accumulated ethanol which is oxidatively metabolized. During the diauxic lag phase in such a system new enzymes which participate in the changed metabolism are synthesized *de novo* and marked increases in lysosomal hydrolases, including aminopeptidases, proteases and RNase are found (Fig. 5.12). These increases in hydrolase activity over

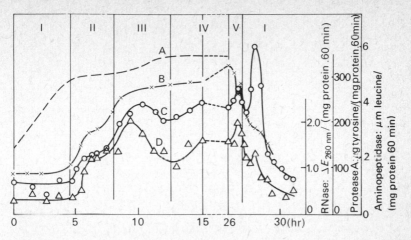

**Fig. 5.12** Diauxic growth of *Saccharomyces cerevisiae* and changes in certain hydrolase activities. The organism was grown in batch culture in a 2 per cent glucose medium at 30° C. Growth phases: (i) first exponential growth phase; (ii) diauxic lag phase; (iii) second exponential growth phase; (iv) stationary phase; (v) first lag phase (of next generation). (A) log dry weight of cells; (B) aminopeptidase; (C) protease A; (D) RNase. The switch from fermentative to oxidative metabolism is accompanied by increased lysosomal enzyme activity in the diauxic lag phase. It is postulated that these lysosomal hydrolases are involved in the breakdown of redundant enzymes at this point. (From Matile, P., Wiemken, A. and Guyer, W. (1971) *Planta (Berl.)*, 96, 43–53. © Springer-Verlag, 1971.)

the relatively lower levels which prevail during exponential growth are believed to be related to the enhanced rate of macromolecular turnover observed in cells in the diauxic lag phase and suggest that a specific function is to hydrolyse the residual enzymes of the glycolytic pathway which become repressed during this period. This system has been examined in detail by Matile, Wiemken and Guyer (1971) who were able to extend their findings to implicate an aminopeptidase, which has a specific lysosomal location in *Saccharomyces cerevisiae*, in the process of protein turnover and biochemical differentiation. In addition they were able to show that an increment of growth, stimulated by a further addition of glucose to starved cultures during the stationary phase, was accompanied by another burst of enhanced hydrolase activity as the redundant enzymes of oxidative metabolism were degraded and those appropriate to fermentative metabolism were synthesized. On the basis of this work it was concluded that the lag phases and stationary growth phases are periods when enzymes, and other macromolecules, that are no longer required are degraded in lysosomes and replaced by a new set of enzymes appropriate to the changed metabolic route.

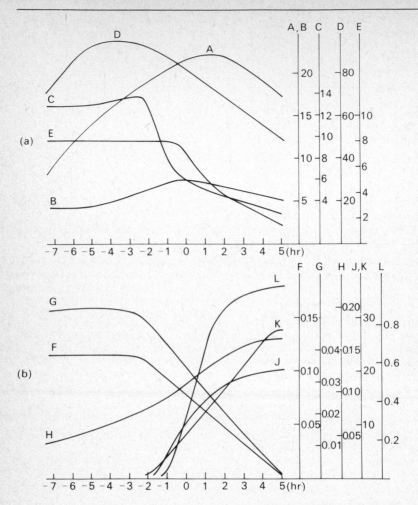

**Fig. 5.13** Changes of some enzyme activities and macromolecule content of cell-free extracts from gills of *Coprinus lagopus* during development of fruiting bodies. Samples were taken at periods from 7.5 hr before release of spores from the autolysing gills (taken as time 0) up to 5 hr after this process.

(*a*) (A) alkaline protease ($\mu$g tyrosine equiv./30 min.); (B) acid protease ($\mu$g tyrosine equiv./30 min.); (C) acid RNase ($\mu$g RNA/60 min.); (D) protein ($\mu$g/mg dry weight of extract). (E) RNA ($\mu$g/mg dry weight). Note the decreases in protein and RNA contents of extracts taken near the time of spore release.

(*b*) (F) aldolase ($\Delta E_{340}$/min.); (G) cytochrome oxidase ($\Delta E_{550}$/min.); (H) acid $\beta$-glucosidase ($\mu$g glucose/10 min.); (J) $N$-acetylglucosaminidase ($\mu$g); (K) acid chitinase ($\mu$g $N$-acetylglucosaminidase (20 min.); (L) acid chitinase (determination of reducing groups/mg $N$-acetylglucosamine equiv./20 min.). Note the precipitous fall in respiratory enzyme activity at spore release accompanied by sharp rises in chitinase activity. It is postulated that chitinase is a lysosomal

enzyme and is responsible for the liquefaction of the gills at spore release. It is not yet clear if other lysosomal hydrolases are involved nor is it known if lysosome disruption causes the death of the cells at this time.
(From Iten, W. and Matile, P. (1970) *J. gen. Microbiol.*, 61, 301—9. By permission of Cambridge University Press.)

### 5.5.2 Lysosomes and Autolysis in the Fruiting Bodies of Coprinus

The genus *Coprinus* contains a number of the familiar 'Ink Cap' fungi. At sporophore maturity there is a wave of spore release progressing from the edges of the gills near the stipe to the periphery of the pileus. Release of spores from the basidia is accompanied by simultaneous autolysis of the gills which results in the familiar inky spore-containing liquid. It has been possible, using special fractionation techniques, to prepare lysosome fractions from gills during the time-course of autolysis (Iten and Matile, 1970). These studies show that the cells of the vegetative mycelium and the fruiting bodies contain acid hydrolases located in vacuoles that have some characteristics of lysosomes. Certain of these enzymes increase in activity during fruit body maturation and decrease in activity soon after the beginning of autolysis of the gills (Fig. 5.13*a*). Difficulties in isolating these vacuoles have so far prevented the elucidation of detailed biochemical knowledge of the events leading to the cellular lytic processes of autolysis, but it is believed that lysosomes may be involved in digestive vacuolation in the mycelium. However, a group of chitinolytic enzymes is located in the vacuoles of the gills and shortly before autolysis there occurs a rapid increase in the specific activity of these enzymes (Fig. 5.13*b*). At the onset of autolysis there is a sharp decline in respiration and in the activity of certain enzymes of the mitochondria. It seems from these observations that the gill cells secrete chitinase from lysosome-like vacuoles to act upon the walls which are then digested with accompanying autolysis. It is not yet known, however, if intracellular release of other suspected lysosomal enzymes is responsible for cell death *before* autolysis of the cell walls.

## Suggestions for further reading

BADENOCH-JONES, P. and BAUM, H. (1973) Progesterone-induced permeability changes in rat liver lysosomes, *Nature, Lond.*, 242, 123—4.

BEAULATON, J. (1967*a*) Localisation d'activités lytiques dans la glande prothoracique du ver à soie du chêne (*Antheraea pernyi* Guér) au stade prénymphal. I. Structures lysosomiques, appareil de Golgi, et ergastoplasme, *J. Microscopie*, 6, 179—200.

BEAULATON, J. (1967*b*) Localisation d'activités lytiques dans la glande prothoracique du ver à soie du chêne (*Antheraea pernyi* Guér.) au stade prénymphal. II. Les vacuoles autolytiques (cytolysomes), *J. Microscopie*, 6, 349—70.

BERJAK, P. (1968) A lysosome-like organelle in the root cap of *Zea mays*, *J. Ultrastruct. Res.*, 23, 233—42.

BLATT, L. M., SCAMAHORN, J. O. and KIM, K. H. (1969) Mechanism of puromycin activation of tadpole liver glycogen synthetase, *Biochim. biophys. Acta*, 177, 553–9.

FILNER, P. and VARNER, J. (1967) A test for *de novo* synthesis of enzymes: density labelling with $H_2O^{18}$ of barley α-amylase induced by gibberellic acid, *Proc. Nat. Acad. Sci. U.S.*, 58, 1520–6.

GAHAN, P. B. (1973) Plant lysosomes, in *Frontiers of Biology*, 29. *Lysosomes in Biology and Pathology*, vol. 3, ch. 5, pp. 69–85, ed. J. T. Dingle. North-Holland Publishing Co., Amsterdam and London.

GAHAN, P. B. and MAPLE, A. J. (1966) The behaviour of lysosome-like particles during cell differentiation, *J. exp. Bot.*, 17, 151–5.

GIBSON, R. A. and PALEG, L. G. (1972) Lysosomal nature of hormonally induced enzymes in wheat aleurone cells, *Biochem. J.*, 128, 367–75.

GREENBAUM, A. L., SLATER, T. F. and WANG, D. Y. (1965) Lysosomal enzyme changes in enforced mammary-gland involution, *Biochem. J.*, 97, 518–22.

HEFTMANN, E. (1971) Lysosomes in tomatoes. *Cytobios*, 3, 129–36.

ITEN, W. and MATILE, P. (1970) Role of chitinase and other lysosomal enzymes of *Coprinus lagopus* in the autolysis of fruiting bodies, *J. gen. Microbiol.*, 61, 301–9.

JONES, R. L. (1973) Gibberellins: their physiological role. *Ann. Rev. Plant Physiol.*, 24, 571–98.

LOCKSHIN, R. A. (1969) Lysosomes in insects, in *Frontiers of Biology* 14A. *Lysosomes in Biology and Pathology*, vol. 1, ch. 13, pp. 363–91, eds. J. T. Dingle and H. B. Fell. North-Holland Publishing Co., Amsterdam and London.

MAHLBERG, P. (1972a) Localization of neutral red in lysosome structures in hair cells of *Tradescantia virginiana*, *Can. J. Bot.*, 50, 857–9.

MAHLBERG, P. (1972b) Further observations on the phenomenon of secondary vacuolation in living cells, *Am. J. Bot.*, 59, 172–9.

MATILE, P. (1969) Plant lysosomes, in *Frontiers of Biology*, 14A. *Lysosomes in Biology and Pathology*, vol. 1, ch. 15, pp. 406–30, eds. J. T. Dingle and H. B. Fell. North-Holland Publishing Co., Amsterdam and London.

MATILE, P., WIEMKEN, A. and GUYER, W. (1971) A lysosomal aminopeptidase isoenzyme in differentiating yeast cells and protoplasts, *Planta (Berl.)*, 96, 43–53.

MATILE, P. and WINKENBACH, F. (1971) Function of lysosomes and lysosomal enzymes in the senescing corolla of the Morning Glory (*Ipomoea purpurea*), *J. exp. Bot.*, 22, 759–77.

PALEG, L. G. (1960) Physiological effects of gibberellic acid. I. On carbohydrate metabolism and amylase activity of barley endosperm, *Plant Physiol.*, 35, 293–9.

PITT, D. (1971) Purification of a ribonuclease from potato tubers and its use as an antigen in the immunochemical assay of this protein following tuber damage, *Planta (Berl.)*, 101, 333–51.

PITT, D. (1973) Solubilization of molecular forms of lysosomal acid phosphatase of *Solanum tuberosum* L. leaves during infection by *Phytophthora infestans* (Mont.) de Bary, *J. gen. Microbiol.*, 77, 117–25.

PITT, D. and GALPIN, M. (1971) Increase in ribonuclease activity following mechanical damage to leaf and tuber tissues of *Solanum tuberosum* L., *Planta (Berl.)*, 101, 317–32.

PITT, D. and GALPIN, M. (1973) Isolation and properties of lysosomes from dark-grown potato shoots, *Planta (Berl.)*, 109, 233–58.

SUTCLIFFE, J. F. and SEXTON, R. (1968) β-Glycerophosphatase and lateral root development, *Nature (Lond.)*, 217, 1285.

SZEGO, C. M., SEELER, B. J., STEADMAN, R. A., HILL, D. F., KIMURA, A. K. and ROBERTS, J. A. (1971) The lysosomal membrane complex. Focal point of primary steroid hormone action, *Biochem. J.*, 123, 523–38.

TATA, J. R. (1966) Requirement for RNA and protein synthesis for induced regression of the tadpole tail in organ culture, *Devel. Biol.*, 13, 77–94.

TATA, J. R. (1971) Hormonal regulation of metamorphosis, *Symp. Soc. Exp. Biol.*, 25, 163—81.

TATA, J. R. (1973) *Metamorphosis*. In series Oxford Biology Readers No. 46, 16 pp. eds. J. J. Head and O. E. Lowenstein. Oxford University Press, London.

TEESENOW, W. (1969) Lytic processes in development of fresh-water sponges, in *Frontiers of Biology*, 14A. *Lysosomes in Biology and Pathology*, vol. 1, ch. 14, pp. 392—405, eds. J. T. Dingle and H. B. Fell. North-Holland Publishing Co., Amsterdam and London.

TIFFON, Y., RASMONT, R., DE VOS, L. and BOUILLON, J. (1973) Digestion in lower metazoa, in *Frontiers of Biology*, 29. *Lysosomes in Biology and Pathology*, vol. 3, ch. 4, pp. 49—68, ed. J. T. Dingle. North-Holland Publishing Co., Amsterdam and London.

WEBER, R. (1964) Ultrastructural changes in regressing tail muscles of *Xenopus* larvae at metamorphosis, *J. Cell. Biol.*, 22, 481 7.

WEBER, R. (1969) Tissue involution and lysosomal enzymes during anuran metamorphosis, in *Frontiers of Biology*, 14B. *Lysosomes in Biology and Pathology*, vol. 2, ch. 15, pp. 437—61, eds. J. T. Dingle and H. B. Fell. North-Holland Publishing Co., Amsterdam and London.

WOESSNER, J. E., Jr. (1965) Acid hydrolases of the rat uterus in relation to pregnancy post-partum involution and collagen breakdown, *Biochem. J.*, 97, 855—66.

YOMO, H. (1960) Studies on the amylase activity substance. IV. On the amylase activity action of gibberellin, *Hakko Kyokaishe*, 18, 600—2.

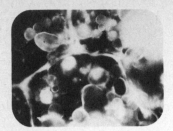

# Chapter 6

# Lysosomes in Disease and Injury

All species show diversities of form and physiology. Within certain limits of variation an organism may be considered normal. At the extremities of the spectrum of diversity a cell, tissue or organism may appear overtly abnormal but the precise borderline between such a condition and normality may be indistinct. Certain abnormal conditions involve harmful deviations from the normal functioning of physiological processes which are manifested as *disease*.

In the previous chapters I sought to define the lysosome system and to indicate its primary role in the processes of intracellular digestion and the relationships with development, senescence and natural autolysis. Soon after the discovery of lysosomes de Duve realized that *in vivo* malfunctioning of this organelle might have serious repercussions for the cell in that leakage of lysosomal hydrolases to the cytoplasm could result in damage to the various cellular components which may lead to disorganization of compartmentation and eventually to cell death. It is possible to envisage several mechanisms whereby lysosomes may be involved in tissue damage and or disease:

(*a*)  Excess of lytic activity due to enhanced and uncontrolled autophagy.

(*b*)  Damage to or permeability changes in lysosome membranes resulting in leakage of enclosed hydrolases to the cytoplasm which in extreme cases could cause cell death.

(*c*)  Uncontrolled or excessive extrusion of lysosomal hydrolases to the exterior of the cell resulting in tissue disorganization and degradation.

(*d*)  An inadequacy of lytic activity such that lysosomes become loaded with indigestible products. This could arise following heterophagic uptake of materials, or to production by the cell of an *abnormal* product which was undegraded by the *normal* complement of lysosomal enzymes. Additionally a genetic abnormality resulting in a deficiency of a lysosomal enzyme would result in the failure of cells to degrade a normal product of metabolism.

The early view that the release of lytic enzymes to the cytoplasm, and perhaps to surrounding tissues, was a primary event resulting in various types of damage has been gradually modified over the years as our knowledge of the complexity of the lysosome system has accumulated, and today more modest claims prevail of the subtle involvement of lysosomes in disease. However, there are several pieces of experimental evidence which suggest that in some circumstances damage to lysosomes may be a primary event which results in discharge of lysosomal hydrolases to the cytoplasm with subsequent cell and tissue damage. I wish now to refer briefly to some of this evidence, frequently derived from tissue and organ culture studies, and then proceed to an examination of the specific role of lysosomes in certain animal and plant diseases.

## 6.1 Experimental Evidence on the Primary Role of Lysosomes in Cell and Tissue Injury

### 6.1.1 Effects of Bacterial Toxins on Lysosomes

We have mentioned earlier (Chapter 3, p. 53) that phagocytic cells, particularly macrophages and leucocytes, form a main line of defence against bacterium and virus infections and that the lysosomes may 'process' antigenic viruses and bacteria. In spite of the animal body defences pathogenic infections often become established although a degree of successful resistance is normal. The classic experiments of Loeffler in 1884 demonstrated that the diphtheria bacilli produced toxins which were disseminated in the blood stream and caused systemic lesions at sites which were found to be free from the bacterium. It is now known that many bacteria produce highly potent toxins which have diverse actions upon organisms. Early studies on lysosomes showed that the streptococcal haemolytic toxins, streptolysins O and S, caused release of enzymes from the lysosomes of granular fractions of rabbit liver and that this preceded liberation of malic dehydrogenase from the mitochondria present in such preparations (Weissman *et al.*, 1963). It seemed from these and other related observations that disruption of lysosomes was an early event in the death of the cell, but subsequent ultrastructural observations revealed that the cell membrane was also affected and this could have preceded labilization of the lysosomes. This group of workers also showed that injection of streptolysin S into the joints of rabbits caused symptoms of extensive acute and chronic arthritis; an observation we shall refer to later on.

### 6.1.2 Effects of Photosensitizing Substances on Lysosome Stability

I have mentioned several times the capacity of secondary lysosomes to accumulate foreign macromolecules by endocytosis. It is also well established that the cells of many organisms are sensitive to specific wavelengths of light in the presence of certain photosensitizing substances such as polybenzenoid hydrocarbons, porphyrins, neutral red and acridine orange which accumulate in the lysosome system. The chemistry of the phenomenon is complex, but it seems that a photon is absorbed by the photosensitizing substance at an appropriate wavelength with an electron becoming excited and subsequent photo-oxidation reactions generating various types of free radicals which cause damage at specific cell sites, including membranes. Experiments by Slater and Riley (1965) showed that the membranes of lysosomes are particularly susceptible to photosensitization damage, possibly caused by the generation of free radicals of the peroxy-type, in the presence of the porphyrin substance, phylloerythrin. Under appropriate wavelengths of light rapid intracellular release of hydrolases occurred from lysosomes which could be potentially damaging to cells. Allison and Young (1969) have reviewed this interesting process and report that sub-lethal as well as lethal damage can occur and the former event may be accompanied by structural abnormalities in the chromosomes which could arise following the specific effect of deoxyribonuclease released from the lysosomes. Allison believes that such effects on the chromosomes arising through this and other causes may implicate lysosome malfunction in some processes leading to carcinogenesis.

### 6.1.3 Effects of Lysosome Contents on Tissues

I have mentioned in some detail the process of extrusion of lysosomal enzymes and its role in the extracellular digestion of connective tissues, and later in this chapter we shall examine the consequences of this phenomenon in arthritic diseases. The damaging effects of the contents derived from purified lysosomes on tissues has been examined in a number of instances. It has been shown experimentally that lysates of purified lysosomes, extracted from rabbit leucocytes, caused acute inflammation at the sites of injection into the skin and joints of rabbits. A course of repeated injections, particularly into joints, caused severe abnormalities in the region of the joints accompanied by extensive cartilage erosion. Furthermore, injections of purified lysosomes seem to cause not only localized effects at the sites of injection but even at areas some distance away.

## 6.2 Lysosomes in some Diseases of Animals

We saw in the last section that there is considerable experimental evidence derived from observations on model systems that labilization or permeability changes of lysosome membranes may cause release of hydrolases to both the cytoplasm and the surrounding tissues. However, in the study of disease processes the tissues affected rarely approximate to model systems and there is no substitute for observations on diseased organisms. Consequently I shall now consider selected examples of the involvement of lysosomes in some diseases of animals and plants.

### 6.2.1 Diseases due to the Accumulation of Indigestible Material within Lysosomes (disorders of lysosome storage)

The normally beneficial process of endocytosis with its involvement in defence and digestion may, on occasions, be harmful. In certain pathological conditions cells can ingest large quantities of indigestible material, either solid or liquid, which cannot be excreted. Much of this material tends to accumulate in the lysosome system to such an extent that it becomes fully laden with the accumulated matter which may eventually be released intracellularly along with the enzymic content of the lysosomes. The cells may then die as a result of lytic activity which may also affect the extracellular environment and result in inflammatory processes. In some cases accumulation in the lysosomes of non-metabolizable materials of cellular origin may be accompanied by *synthesis* and ultimate *release* of large quantities of lysosomal enzymes to the cytoplasm and eventually to the surrounding tissues. Natural products may also accumulate in lysosomes as a normal event, e.g. lipofuscin in aging tissues, also certain metabolic products are known to accumulate during autophagy and remain undegraded because of some inborn error of lysosome metabolism. On occasions the accumulating material may itself be an abnormal metabolic product which is undegradable by the *normal* complement of lysosomal hydrolases.

*6.2.1.1 Silicosis and Asbestosis.* Much of the foreign matter which is inhaled into the lungs is eventually taken up by alveolar macrophages where it may remain almost indefinitely in a harmless state. However, certain types of dust, especially silica and asbestos, stimulate fibrogenic reactions associated, respectively, with *silicosis* and *asbestosis*. Inhaled silica particles are phagocytosed by alveolar macrophages and soon become incorporated into secondary lysosomes via the phagosomes. It is believed (Allison, 1969; 1970) that the toxic properties of silica particles are related to their capacity to damage lysosome membranes through hydrogen bonding interactions. Some evidence for this belief is derived from experiments with poly-2-vinylpyridine-N-oxide (PPNO) which is readily taken up into phagosomes and which reduces the toxic effects of silica. It was proposed by Allison and co-workers that PPNO, which possesses *N*-oxide groups which readily form hydrogen bonds with silicic acid, preferentially reacts with the silicic acid on the surface of the silica particles in the lysosomes so maintaining the integrity of the lysosome membrane. Under conditions of excessive exposure to silica particles a chain reaction could be initiated wherein damaged lysosomes release both hydrolases and silica particles to the cytoplasm which could, in theory at least, lead to death of the macrophages. The silica particles so released could be taken up by further macrophages and the cycle of events repeated. As these processes take place fibroblasts in the affected area are stimulated to produce collagen during fibrogenesis.

Asbestos fibres may likewise be taken up by the tissues of the lungs with accompanying macrophage killing and fibrogenic effects which could result in the symptoms of asbestosis. There is little information available on the manner in which lysosome membranes are affected by asbestos particles nor indeed is there certain knowledge that asbestos in the lysosomes is a cause of asbestosis, although such material is known to be membrane-active. An additional effect of inhalation of some types of asbestos, silica, metal powders and certain other particles is the development of tumours of the lung. Such an abnormality was produced by Rajan *et al.* (1972) in explants of lung tissue cultured in the presence of asbestos fibres, but the mechanisms of these secondary effects in relation to lysosome function are obscure.

*6.2.1.2 Gout.* A possible relationship between crystal deposits and gout has been suspected since the last century. Attacks of gout and pseudo-gout have been experimentally induced by injecting microcrystalline suspensions of urate and calcium pyrophosphate, respectively, into volunteers. Conversely, these substances administered in solution do not cause symptoms of gout. Furthermore colchicine which is used in the therapy of gout is also known to restrict phagocytosis and degranulation (secondary lysosome formation) by polymorphonuclear leucocytes. This sort of evidence implies that the crystals of urate and pyrophosphate associated with gout and pseudo-gout are taken up in phagocytosis by polymorphs. The uptake of crystals into endocytic vacuoles and the degranulation may be accompanied by some limited leakage of lysosomal enzymes into the cytoplasm and to the exterior of the cell and thereby be responsible for inflammation and in addition account for some of the cartilage erosion which accompanies these diseases. Although lysosomes seem to be primarily implicated in such disorders it is more than likely that the current ideas regarding the physiological basis of gout are over-simplified.

### 6.2.2 Storage Diseases due to Genetic Abnormalities of Lysosomes

A number of inherited storage disorders of human beings are known in which polysaccharides, lipids and other materials accumulate in various tissues. The discovery of a key role of lysosomes in the intracellular digestive system permitted speculation that inherited storage diseases may be related to a congenital defect in or absence of a specific lysosomal acid hydrolase. Thereby developed the concept of inborn lysosomal diseases in which *normal* cell macromolecules remained undigested and accumulated in secondary lysosomes following entry to the vacuolar system during endocytosis or, endogenously, via autophagy. In such disorders, affected subjects might appear almost normal at birth with the symptoms developing gradually with time following the progressive accumulation of undegradable residues. Indeed, this is often the normal pattern of symptom expression in these diseases which are frequently fatal in childhood.

*6.2.2.1 Pompe's Disease* (Type II glycogenosis). Several forms of human pathological conditions exist, known collectively as glycogenoses, in which glycogen accumulates in excessive amounts in most tissues, especially the heart, liver and skeletal muscles. Type II glycogenosis is a particular form of this condition in which there is no impairment of the mechanism responsible for the phosphorolytic breakdown of glycogen based in the cytoplasm, but owing to a defect in the lysosomes glycogen accumulating in autophagic bodies, which contain no active phosphorolytic enzymes, remains unmetabolized and gradually overloads the vacuoles and the cells. The specific lysosomal defect is in respect of $\alpha$-1,4-glucosidase (acid maltase) which is capable of hydrolysing maltose, linear oligosaccharides and the branched chains of glycogen to glucose. The lysosomes become so overloaded that glycogen also accumulates in massive quantities in the cytoplasm, but this fact alone is not responsible for serious tissue dysfunction. Death of the subject results from the degeneration of muscle fibres by cathepsins which are released to the cytoplasm when the lysosomes disrupt through mechanical forces resulting from overloading with glycogen (Hers and Van Hoof, 1969; 1973).

*6.2.2.2 Some other Genetic Disorders of Lysosomes.* Hurler's syndrome (gargoylism) is also rare and is characterized by dwarfism, deformity and mental retardation. It is usually fatal and generally confined to unfortunate children who take on the familiar appearance of Tweedledum and Tweedledee from *Alice through the Looking Glass*, characters who must surely have been based on subjects displaying symptoms of gargoylism who were known to Lewis Carroll. Physiological symptoms include accumulation of mucopolysaccharides in certain tissues and excretion of high levels of these substances in the urine. There is uncertainty regarding the precise metabolic basis for Hurler's disease and for some of the closely related disorders. Ultrastructural observations reveal large quantities of stored mucopolysaccharides and glycolipids in enlarged lysosomes of various tissues including liver, nerve and brain. Biochemical work shows a variable pattern of enzyme activity in tissues of different patients such that it has been difficult to pinpoint in all but a few cases the precise enzyme deficiency responsible for the condition. It has emerged, however, that a deficiency of lysosomal $\alpha$-fucosidase which hydrolyses fucoside

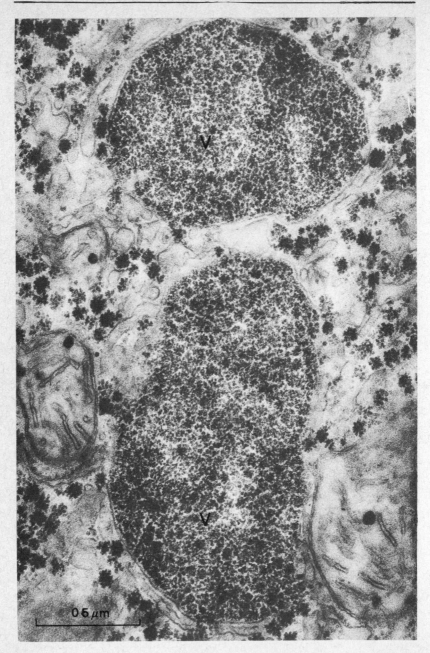

**Fig. 6.1** Type II glycogenosis (Pompe's disease). Thin section of a portion of a

human liver parenchyma cell. The enlarged vacuoles (V) are filled with glycogen. These structures are secondary lysosomes which have accumulated glycogen during autophagy. (From Baudhuin, P., Hers, H. G. and Loeb, H. (1964) *Lab. Invest.*, 13, 1139—52. By courtesy of The Williams and Wilkins Company, © 1964. Original print supplied by Professor P. Baudhuin.)

moieties of various macromolecules, appears to be primarily responsible for certain types of gargoylism. Partial deficiencies of β-galactosidase have also been found in patients displaying Hurler syndrome, but further work is necessary to define more closely the metabolic bases of the various forms of this disease.

A number of other inborn errors of metabolism have recently been examined and information is accumulating (Hers and Van Hoof, 1969; 1973) which permits some extension of the concept of genetic abnormalities of lysosomes to some of these disorders. The nature of these lysosomal disorders is such that theoretically it ought to be possible to treat them by replacement therapy consisting of administration of intravenous injections of the purified enzyme that is deficient which should then enter the lysosome system by endocytosis. Indeed the *cellular* symptoms of Pompe's disease have been alleviated when vacuolar glycogen was found to disappear in a patient treated with α-glucosidases extracted from fungi. Unfortunately the child died from immunological complications arising through the presence of the foreign protein. It would seem necessary to use for this purpose purified enzyme derived from readily available human sources, e.g. the placenta, in order to overcome the dangers of adverse immunological reactions, or to enclose the enzymes in artificial membranes (liposomes) which do not elicit antibody formation after administration and which fuse with lysosomes (Gregoriadis, 1973).

### 6.2.3 Lysosomes in Arthritis
The ends of the long bones of a typical healthy joint are covered by smooth layers of *articular cartilage* with the joint partially lined and surrounded by the *synovial tissue*. The enclosed space within the joint contains *synovial fluid* which acts as a lubricant. During rheumatoid arthritis the synovial tissues become inflamed and in later stages erosion of the articular cartilage occurs such that friction develops in the joint which then becomes painful to move and may gradually cease to function. The cause of the initial inflammation is not known with certainty although there is evidence that either immunological reactions or microbial infections, or both, may be involved. In addition it is not clear if cartilage erosion arises from activities of the synovial cells or from those of the cartilage.

Cartilage consists of special cells embedded in an extracellular matrix made up of the protein *collagen* and a protein—polysaccharide complex known as *proteoglycan*. The cells in cartilage remain active and are responsible for the continuous turnover of the components of the connective tissues. Histochemical and biochemical analyses reveal that arthritic joints are characterized by a deficiency in the proteoglycan component of the matrix and that this arises through the rapid breakdown rather than the reduced synthesis of this component. It was established early in the studies on lysosomes that rheumatoid synovial fluid contained a higher than normal level of lysosomal enzymes including cathepsin, acid phosphatase and β-galactosidase when considerable speculation developed concerning the origin of these enzymes. I have already mentioned the observations of Mellanby and Fell on the effects of excess vitamin A on the breakdown of cartilage of limb-bone rudiments in organ culture (Chapter 4, p. 85) and the accompanying synthesis and secretion of cathepsin D and other lysosomal enzymes during the process. These observations prompted the suggestion by the Cambridge group that degradation of articular cartilage in arthritis may be due to the release of lysosomal enzymes from cells in the region of the cartilage. Extensive evidence supporting this view is now available and includes cytochemical observations revealing the increased presence of lysosomes in rheumatoid synovial tissues, and certain physical changes within lysosomes in these tissues which are indicative of increased lysosome lability. Ultrastructurally rheumatoid synovial tissues show enhanced phagocytosis and autophagy. It is also well established that injection of the contents of lysosomes into joints can induce inflammation along with several of the other symptoms associated with rheumatoid arthritis. Furthermore the concept of the 'two-stage' digestion of extracellular matrix by lysosomal enzymes extruded to the microenvironment of the cell has received strong support from a number of investigations. Until recently, however, it has not been possible to identify precisely the proteolytic enzyme(s) involved in cartilage degradation. The Cambridge group has now managed to purify a cathepsin D from chick liver and found that a specific antiserum to the enzyme is a potent inhibitor of the cathepsin D activity of chick bone rudiments. The key observation was also made that resorption of chick limb-bone cartilage in culture was greatly inhibited by the presence of the anti-cathepsin D serum. It is likely, therefore, that lysosomal cathepsin D plays an important role in extracellular matrix resorption and extracellular release of such an enzyme within the region of articular cartilage would, in theory, have a devastating effect upon the functioning of the joint. The use of fluorescent-labelled anti-cathepsin D serum coupled with ultrastructural observations using serum labelled with electron-dense compounds may throw further light on the changes in distribution of cathepsin D in diseased joints.

Thus, the evidence for the involvement of lysosomes in arthritis is convincing, and it may also be that products of extracellular digestion are phagocytosed and perhaps processed as 'foreign' antigenic material within cells so generating an autoimmune response within the region of the diseased joint. The present belief is that several diverse factors may initiate primary responses in inflammatory joint diseases and that the lysosome system may be involved at various stages in these disorders. It is particularly interesting that several agents which have found use in the therapy of joint disorders, e.g. hydrocortisone, colchicine and gold salts, have direct effects on the functioning and stability of the lysosome system.

The literature on the known and speculative roles of lysosomes in animal pathology continues to expand rapidly. The present treatment is necessarily brief, access to the current research literature is provided by the recent article of Slater (1974).

## 6.3 Lysosomes and Plant Disease

One of the earliest manifestations of disease in plants is the drastic alteration in the nature of cell membranes, frequently measured experimentally as electrolyte release from the cells of infected areas or as changes in their response to vital dyes. There are many factors known to mediate membrane changes in plant disease including a variety of enzymes and toxins, pH, free radicals and oxidizing systems. The bases of action of these is largely unknown, but one possibility is that they could interfere with the fine controls regulating membrane fusion and cellular compartmentation, and evidence is available that some disease situations in plants, as in animals, involve interference with the lysosome system.

### 6.3.1 Evidence for the Release of Lysosomal Enzymes from a Bound Form in Plant Diseases caused by Pathogenic Fungi

Considerable information has accumulated concerning the nature of acid hydrolases in fungus infections of plant tissues. During infections of bean leaves caused by a particular rust fungus there is increased overall activity of acid phosphatase in the host: parasite complex owing to contributions of this enzyme by *both the host and the parasite*. A similar situation exists in the necrotrophic infection of potato tubers caused by the pink rot fungus, *Phytophthora erythroseptica*, when increased activities of acid phosphatase, ribonuclease and carboxylic esterases occur in the diseased tissues. Since both the pathogen and the host contain acid hydrolases compartmentalized within their respective lysosome systems, several intriguing possibilities could occur during infection which might account for these increases in total hydrolase activity. I have already mentioned the extensive capacity of fungi to secrete hydrolases to the exterior. These hydrolases, including those of known lysosomal location, are of importance to the parasite since they degrade complex host materials into simpler utilizable metabolites. It is also possible that hydrolytic enzymes of host origin, e.g. proteases, glucosidases and phospholipases could degrade enzymes, toxins and other metabolites secreted by the pathogen.

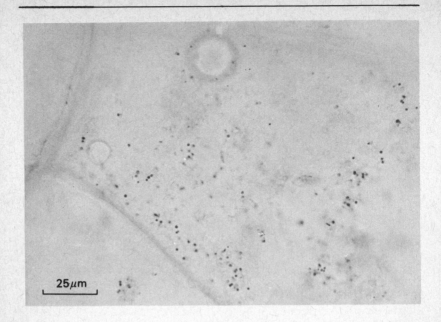

Fig. 6.2    Lysosomes of potato tubers. Section of parenchyma cell of potato tuber fixed in formol-calcium and stained for acid phosphatase by the standard coupling azo dye method of Grogg and Pearse, 1952. Acid phosphatase activity is mainly confined to the lysosomes. (From Pitt, D. and Coombes, C. (1968) *J. gen. Microbiol.*, 53, 197–204. By permission of Cambridge University Press.)

Furthermore it seems a possibility that acid hydrolases of the host could be decompartmentalized and lead to cell death and autolysis of both host and parasite. Recent work on the role of lysosomes in plant host: fungus parasite interaction has started to unravel some of these problems and provides partial answers to questions on the role of lysosomes in plant disease.

Using histochemical methods Pitt and Coombes (1968) demonstrated the existence in potato tuber tissues of sub-cellular particles rich in lysosomal hydrolases (Fig. 6.2), and went on to show that these fulfilled a number of the criteria for lysosomes. Infection of potato tubers by *Phytophthora erythroseptica* was accompanied by swelling and disruption of host lysosomes as revealed in tissue sections stained histochemically for acid hydrolase activity. Widespread cytoplasmic staining of acid phosphatase was also a characteristic feature not only of this disease (Fig. 6.3), but also of tubers infected with *Phytophthora infestans* (causing potato tuber blight) and *Fusarium caeruleum* (responsible for dry-rot disease of potato tubers). Attempts to seek biochemical support for these findings

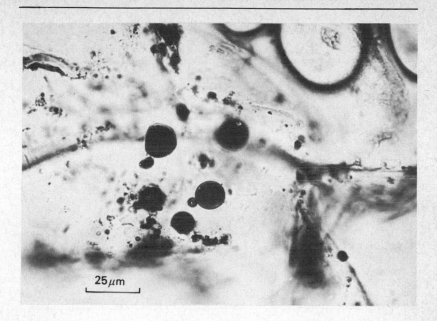

**Fig. 6.3** Effect of fungus infection on potato tuber lysosomes. Section of parenchyma cell of a potato tuber infected with the agent of pink rot disease, *Phytophthora erythroseptica*. Stained for acid phosphatase activity as in Fig. 6.2. Note the heavy diffuse cytoplasmic staining of acid phosphatase, enlargement and disruption of particulate sites of enzyme activity. (From Pitt, D. and Coombes, C. (1968) *J. gen. Microbiol.*, 53, 197—204. By permission of Cambridge University Press.)

through isolation from tubers of particulate hydrolases have been unsuccessful so far owing to a combination of difficulties in homogenizing this tissue and to biochemical problems probably related to the high endogenous activities in tubers of phospholipases and lipoxygenases. However, histochemical work on potato tuber callus grown in tissue culture revealed similar behaviour of particle-bound hydrolases during infections by these pathogens. Furthermore the ease with which such tissue can be homogenized permitted its use in biochemical studies which showed that decreased activity of acid phosphatase and RNase in the particulate fraction of infected callus cells was accompanied by increased activities of these enzymes in the supernatant fluid fractions (Table 6.1). An interesting additional observation in this work was the consistent excess recovery of RNase from homogenates of callus cells infected with *P. infestans* and *F. caeruleum*. Further investigation of this particular point (Pitt and Galpin, 1973*a*) by use of an immunochemical assay for measuring enzyme *protein* in addition to determinations of enzyme *activity* suggested that

**Table 6.1**   Acid phosphomonoesterase activity in particulate (P) and supernatant fluid (S) fractions of potato callus tissue using $\beta$-glycerophosphate as substrate (after Pitt and Coombes, 1969).

| Treatment | Activity of fractions Inorganic phosphate released ($\mu$g $P_i$/ 20 min./mg protein) | | Total activity (P + S fractions) in fractions (%) | |
|---|---|---|---|---|
| | P | S | P | S |
| *P. erythroseptica* | | | | |
| Uninfected | 3.07 | 2.30 | 57.19 | 42.81 |
| 4-Day infected | 1.04 | 2.31 | 31.01 | 68.99 |
| *P. infestans* | | | | |
| Uninfected | 5.26 | 3.90 | 57.42 | 42.58 |
| 5½-Day infected | 2.60 | 8.97 | 22.48 | 77.52 |
| *F. caeruleum* | | | | |
| Uninfected | 3.06 | 1.43 | 68.11 | 31.89 |
| 4½-Day infected | 1.99 | 4.88 | 28.94 | 71.06 |

the increased activity of RNase associated with infections by *F. caeruleum* may be due to a similar biochemical mechanism as occurs following mechanical damage to potato tubers and leaves and may involve, primarily, the activation of pre-formed enzyme rather than *de novo* synthesis of enzyme protein. Current evidence from the author's laboratory tends to support the view that increased RNase activity occurring in infection and mechanical damage may be located both in the lysosomes and the soluble phase of cell extracts.

Although the results outlined above revealed that infection was accompanied by decreased particulate activity and increased activity of enzyme in the soluble phase of cell homogenates this may not necessarily have arisen through a release of enzyme from the lysosomes. It was possible that infection resulted in specific inactivation of particulate hydrolases by an inhibitor of either fungus or host origin. Additionally, it is known that fungi contain particulate hydrolases which may be secreted and add to the soluble enzyme activity of the diseased system. The key question was whether the enzyme associated with the host cell lysosomes was actually released and if so could it be detected within the soluble phase of infected tissues. To answer this question it was necessary to devise a method for isolation and characterization of lysosomes, preferably from a system which was not unduly complicated by the formation on infection of novel forms of the hydrolases under examination. Using the technique developed for the improved isolation of lysosomes from potato sprouts and the application of phospholipase digestion to release fully and

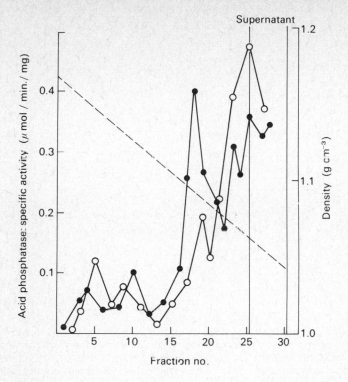

**Fig. 6.4** Changes in the location of acid phosphatase (*p*-nitrophenyl phosphate substrate) in homogenates of potato leaves during infection by *Phytophthora infestans* (potato blight pathogen). Distribution of acid phosphatase activity in Ficoll density gradients after ultracentrifugation of homogenates from normal leaves (•————•) and infected leaves (o————o); —————, density. Note the decrease in activity of particulate enzyme (fraction 20) following infection and the increase in the activity in the soluble phase (fraction 25). (From Pitt, D. (1973) *J. gen. Microbiol.*, 77, 117—25. By permission of Cambridge University Press.)

allow characterization of the complements of soluble and particulate hydrolases (Pitt and Galpin, 1973*b*) it has been possible to follow the fate of lysosomal hydrolases during infection of potato leaves in potato late blight infections caused by *Phytophthora infestans*. These experiments clearly showed that infection resulted in reduced particulate activity of acid phosphatase accompanied by increased activity in the soluble phase of the cell (Figs. 6.4 and 6.5). The same criticism applies to these findings as to the earlier ones that possible inactivation of particulate enzyme and/or secretion of acid phosphatase by the pathogen may be taking place. However, it was found that the potato leaf lysosomes contained a particular

139

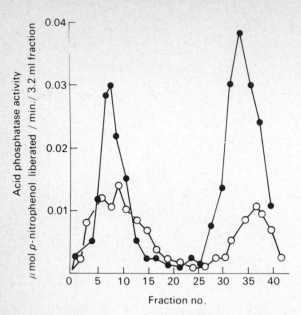

**Fig. 6.5** Changes in the particulate location of molecular forms of acid phosphatase in potato leaves during potato blight caused by *Phytophthora infestans*. Distribution of molecular forms of acid phosphatase (*p*-nitrophenyl phosphate substrate) resolved by gel filtration on Sephadex G-100 in the particulate fraction (solubilized by snake venom digestion after separation of lysosomes) of normal potato leaves (•————•) and leaves infected with *Phytophthora* (o————o). Note the decreased activity of particle bound activity which follows infection. (From Pitt, D. (1973) *J. gen. Microbiol.*, **77**, 117—25. By permission of Cambridge University Press.)

form of acid phosphatase, detectable by gel filtration on Sephadex G-100, not found in the soluble phase of uninfected leaves and whereas this component of the particulate fraction of leaves decreased on infection its content in the supernatant fluid fraction of leaf homogenates (i.e. the soluble phase) simultaneously increased (Fig. 6.6). Since the pathogen lacked the capacity to produce this molecular form of acid phosphatase in culture it seems likely that it failed to do so during infection especially as this enzyme showed no net increases over levels of activity in the controls. On the basis of these experiments it is not possible to exclude either that extralysosomal phosphatase may be synthesized *de novo* in infected tissue or that increased fragility of lysosomes arising as a result of disease makes them more liable to release their enzymic contents during homogenization. Although none of these possibilities can be excluded it is a reasonable assumption on the basis of the various histochemical and biochemical evidence that infection of potato tissues may result in a

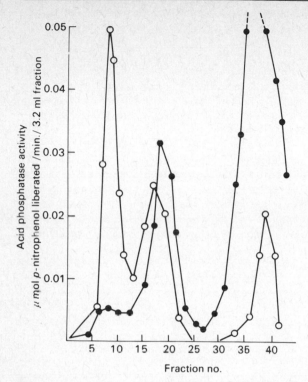

Fig. 6.6   Liberation of acid phosphatase from particulate sites in potato leaves during infection by *Phytophthora infestans*. Molecular forms of acid phosphatase (*p*-nitrophenyl phosphate as substrate) resolved by gel filtration on Sephadex G-100 of snake venom digests of the supernatant fluid fractions (soluble enzymes) from healthy leaves (● ——————— ●), and infected leaves (○ ——————— ○). Note the large peak of activity of the high molecular weight form of acid phosphatase (fraction 10) occurring in the soluble phase of homogenates derived from infected potato leaves compared to the negligible quantity of this form in healthy leaves. During infection this form is released from the lysosomes, cf. Fig. 6.5. (From Pitt, D. (1973) *J. gen. Microbiol.*, 77, 117—25. By permission of Cambridge University Press.)

redistribution of acid phosphatase activity and may involve a release to the soluble phase of the cell of those molecular forms of this enzyme that are characteristically membrane-bounded in healthy cells. Those factors responsible for such release *in vivo* are so far undefined.

Thus, there seems reasonable evidence for the involvement of lysosomes in necrotrophic infections of plants especially some soft-rots and related disorders. The belief has recently been expressed by Pitt and Galpin (1973*a*) and Wilson (1973) that lysosomes may be involved in resistance responses of plants to infection, particularly in the hypersensitive reaction.

The hypersensitive response is a resistance reaction in plants whereby necrotic areas are formed at sites of infection where the pathogen is inactivated or killed. Although this is perhaps the most frequently encountered form of resistance to disease found in plants the biochemical mechanism of the response is largely obscure. It is an attractive idea that the extreme host susceptibility to infection, which forms the basis of the response, is characterized by a localized tissue death involving lysosome disruption. This area of research is amongst the most challenging and potentially one of the most rewarding in plant pathology.

### 6.3.2 Possible Role of Lysosomes in Other Types of Plant Disease

So far we have considered only the necrotrophic response during plant infection which is characterized by the development of necrotic spots or lesions. No information is yet available on the role of lysosomes in biotrophic infections except for that already briefly mentioned on hydrolase activity in rust infections. There is little evidence at present on the involvement of lysosomes in plant virus diseases. Limited observations have been made on the changes in activity and distribution of lysosomal acid phosphatase and esterase during the induction of crown gall in tomato plants caused by *Agrobacterium tumefaciens*. The exact role of increased hydrolases in crown gall disease is unknown, but work that has been done suggests that enhanced lysosomal and cytoplasmic activity of hydrolases need not lead to cell death and in this case may be involved in a so far unspecified manner in tissue hypertrophy (overgrowth).

### 6.3.3 Lysosomes in Mycorrhizal Associations

It is now becoming increasingly apparent that the roots of many higher plants form *mycorrhizal associations* with fungus partners from which both components derive mutual benefits. The fungal invasion may be largely confined to the outer layers of the root where a surface weft of fungus covers the young roots. This type of symbiotic association is known as an *ectotrophic mycorrhiza* in which the fungus derives carbohydrate from the plant and the latter benefits from the facility of the fungus to absorb additional water and mineral salts. Recent observations (Bartlett and Lewis, 1973) have revealed the presence of high activities of surface phosphatases on mycorrhizal roots of beech which may indicate a capacity of the fungus partner to mobilize phosphates that are generally in a form unavailable to roots lacking the association. The exact role of the lysosome system of the fungus in the process of phosphate uptake is obscure, but these observations suggest both a secretory and absorptive function. This is an area of current research which may provide an explanation for the metabolic basis of the functional significance of the fungus partner.

In addition fungi form *endotrophic* mycorrhizal associations with certain plants, e.g. heaths and orchids. Such associations are characterized

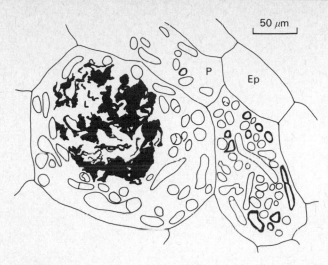

50 μm

P

Ep

**Fig. 6.7** Schematic diagram of cells of a protocorm of *Dactylorhiza purpurella* infected by the fungus *Thanatephorus cucumeris* (*Rhizoctonia solani*). Ep, uninfected epidermal cell; P, peripheral parenchyma infected by thick or thin-walled hyphae; L, lysed fungus surrounded by hyphae of second infection in central parenchyma. (From Williamson, B. (1973) *Planta* (*Berl.*), 112, 149—58. © Springer-Verlag, 1973.)

by a largely internal location of the fungus partner within the cortical parenchyma cells of the root, but some direct hyphal contact with the soil is also retained. The intracellular hyphae undergo lysis and seem to provide nutrients for the higher plant whilst in addition the fungus partner can translocate carbon and phosphate into the host from external sources (Fig. 6.7). A dynamic balance seems to exist whereby the aggressive spread of the infectious fungus partner is kept in check by the plant. For many years mycologists have been fascinated by the mechanism controlling aggressive spread of the fungus hyphae in such associations and some workers consider these mycorrhizas display a form of mutually beneficial *balanced parasitism*. It has been generally assumed that digestive enzymes are produced by the host which then cause lysis of the fungus partner. Undoubtedly root cells have the digestive capacity to carry out such a process (Fig. 6.8), but it is also possible that the fungus (the endophyte) under the circumstances in which it exists may undergo autolysis following the intracellular release of its own lysosomal hydrolases. A recent attempt has been made by Williamson (1973) to solve the problem of whether the host, the endophyte or both partners synthesize hydrolases which are ultimately involved in lysis of the fungus. Careful cytochemical examinations were made of the early changes in location of acid phosphatase and

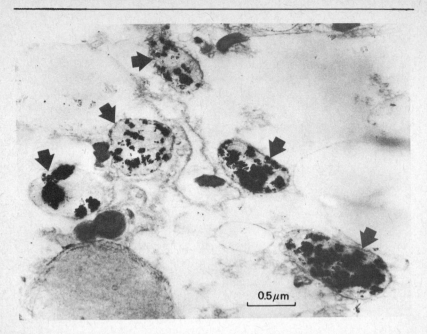

0.5μm

**Fig. 6.8**   Ultrastructural localization of acid phosphatase in orchid protocorm. Electron micrograph showing localization of acid β-glycerophosphatase activity in organelles bounded by a single membrane (arrows) in a parenchyma cell of the orchid (*Dactylorhiza purpurella*) symbiotically infected by *Rhizoctonia solani*. It is believed that these structures are lysosomes which may be involved in lysis of the endophyte. (Original unpublished print supplied by Dr B. Williamson.)

esterase in the protocorms of the orchid *Dactylorhiza purpurella* infected by the endophyte *Thanatephorus cucumeris*, and although it is not yet possible to decide if endophytic hyphae lyse autolytically or are digested by the enzymes of the host it is apparent that membrane changes occur in the lysosome systems of both partners and the suggestion was made that the apparent release of acid hydrolases in the host and fungus cells may be related to the depletion of the phosphate ester pool in cells which have supported extensive hyphal growth. The net effect of lysis would be to make phosphate available to surviving host cells subsequent to phosphatase-mediated hydrolysis of cellular components.

Although the above account summarizes the considerable evidence that malfunctioning of the lysosome system may be a component of disease situations it is not yet clear in any instance of plant or animal disease if lysosome disorganization is a cause or a consequence of cell death. Only through increased understanding of the nature of cell death itself shall we be able to evaluate more fully the role of lysosomes in disease and injury.

## Suggestions for further reading

ALLISON, A. C. (1970) Effects of particles on lysosomes, *Brit. Assoc. for the Adv. of Sci.*, 27, 137—40.

ALLISON, A. C. and YOUNG, M. R. (1969) Vital staining and fluorescence microscopy of lysosomes, in *Frontiers of Biology*, 14B. *Lysosomes in Biology and Pathology*, vol. 2, ch. 22, pp. 600—28, eds. J. T. Dingle and H. B. Fell. North-Holland Publishing Co., Amsterdam and London.

BARRETT, A. J. (1970) The role of lysosomes in arthritis, *Brit. Assoc. for the Adv. of Sci.*, 27, 140—3.

BAUDHUIN, P., HERS, H. G. and LOEB, H. (1964) An electron microscopic and biochemical study of Type II glycogenosis, *Lab. Invest.*, 13, 1139—52.

BARTLETT, E. M. and LEWIS, D. H. (1973) Surface phosphatase activity of mycorrhizal roots of beech, *Soil Biol. Biochem.*, 5, 249—57.

FELL, H. B. (1970) The chain of discovery, *Brit. Assoc. for the Adv. Sci.*, 27, 129—37.

GREGORIADIS, G. (1973) Molecular Trojan horses, *New Scientist*, 60, 890—3.

HERS, H. G. and VAN HOOF, F. (1973) *Lysosomes and Storage Disorders.* Academic Press, New York and London.

HISLOP, E. C. and PITT, D. (1974) Sub-cellular organization and plant host: parasite interaction, *Physiological Plant Pathology*, Encyclopedia of Plant Physiology, New Series, Springer-Verlag, Berlin. (In press.)

PITT, D. (1973) Solubilization of molecular forms of lysosomal acid phosphatase of potato leaves during infection by *Phytophthora infestans.* (Mont.) de Bary, *J. gen. Microbiol.*, 77, 117—25.

PITT, D. and COOMBES, C. (1968) The disruption of lysosome-like particles of *Solanum tuberosum* cells during infection by *Phytophthora erythroseptica* Pethybr, *J. gen. Microbiol.*, 53, 197—204.

PITT, D. and COOMBES, C. (1969) Release of hydrolytic enzymes from cytoplasmic particles of *Solanum tuberosum* tissues during infection by tuber-rotting fungi, *J. gen. Microbiol.*, 56, 321—9.

PITT, D. and GALPIN, M. (1973*a*) Role of lysosomal enzymes in pathogenicity, in *Fungal Pathogenicity and the Plant's Response*, 3rd Long Ashton Symposium, eds. R. J. W. Byrde and C. V. Cutting. Academic Press, London and New York.

PITT, D. and GALPIN, M. (1973*b*) Isolation and properties of lysosomes from dark-grown potato shoots, *Planta (Berl.)*, 109, 233—58.

RAJAN, K. T., WAGNER, J. C. and EVANS, P. H. (1972) The response of human pleura in organ culture to asbestos, *Nature, Lond.*, 238, 346—7.

SLATER, T. F. (1974) Lysosomes (with a short note on peroxisomes), in *Companion to Biochemistry*, ch. 17, pp. 511—51, eds. A. T. Bull, J. R. Lagnado, J. O. Thomas and K. F. Tipton. Longman, London.

SLATER, T. F. and RILEY, P. A. (1966) Photosensitization and lysosomal damage, *Nature, Lond.*, 209, 151—4.

WEISSMAN, G., KAISER, H. and BERNHEIMER, A. W. (1963) Studies on lysosomes III. The effect of streptolysins O and S on the release of acid hydrolases from a granular fraction of rabbit liver, *J. exp. Med.*, 118, 205.

WILSON, C. L. (1973) A lysosomal concept for plant pathology, *Ann. Rev. Phytopath.*, 11, 247—72.

WILLIAMSON, B. (1973) Acid phosphatase and esterase activity in orchid mycorrhiza, *Planta (Berl.)*, 112, 149—58.

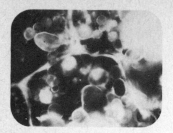

# Methods for Studying Lysosomes

This section summarizes briefly some elementary techniques used in lysosome studies. The majority of these have been used at some time in the author's laboratory and most of them are of a basic nature requiring equipment which is generally available and should therefore be within the scope of most undergraduate students with interests in cell physiology. Space does not permit a full treatment of all the techniques mentioned, but most are accompanied by references which permit access to the appropriate literature.

## Microscopy

### 1. Histochemistry

Several good methods are available for the sub-cellular localization of lysosomal enzymes since fortunately these hydrolases are amongst the few enzymes that withstand fixation. Methods for acid phosphatase, $\beta$-galactosidase, $\beta$-glucuronidase, sulphatases and various esterases have been very successful at the level of light microscopy and in recent years some success has been achieved in modifying and extending these procedures in studies at the level of ultrastructure.

(*a*) **Methods for acid phosphatase.** (i) *The lead salt procedure*. Originally introduced by Gomori in 1941 this method revolutionized histochemistry and heralded the upsurge of interest in sub-cellular localization of enzymes by histochemical techniques. Essentially the method involves the reaction

146

of phosphate, liberated by acid phosphatase from the organic phosphate esters serving as substrates, with a lead salt resulting in the formation of lead phosphate. These deposits, generally occurring at the sites of enzyme activity, are visualized by converting the lead phosphate to black sulphide following treatment with ammonium sulphide.

### Modified Gomori procedure for acid phosphatase

Fix tissue blocks (5 mm diameter) in cold neutral formalin or formol-calcium for 16 hr at $0°-4°$ C. Alternatively fresh frozen sections or coverslip cultures (in the case of tissue cultures or protists) may be fixed.

Wash blocks/sections in 0.88 M-sucrose (containing 1 per cent, w/v, gum acacia) for 24 hr at $0°-4°$ C.

Wash in distilled water for a few minutes.

Section on a cryostat (as applicable).

Pick up sections on slides (these may be subbed in 1 per cent, w/v, gelatin).

Incubate material at $37°$ C in *freshly prepared* 0.01 M-sodium $\beta$-glycero-phosphate in 0.05 M-acetate buffer, pH 5.0, containing 0.004 M-lead nitrate. Incubation times vary with the tissue, e.g. kidney, liver 10—20 min.; plant material and fungi *c*. 30 min.; protozoa 1 hr. Some tissues require up to 3—4 hr.

Wash in several changes of distilled water.

Mount in glycerine jelly.

Sites of enzyme activity appear black.

Some of the caprice associated with the original procedure has been eliminated by minor modifications, but the method still gives variable and often inexplicable results. This is particularly so with plant materials which may contain high levels of endogenous phosphates. Extensive controls are necessary and should include incubations omitting substrate and lead and containing NaF (0.01 M).

(ii) *Azo dye methods for acid phosphatase*. In 1949 Seligman devised a novel procedure for the histochemical demonstration of acid phosphatase in which naphthol, liberated from naphthol esters, was coupled with diazonium salts to give an azo dye at the site of enzyme activity. The majority of modern studies on the histochemistry of lysosomal hydrolases depend heavily on this principle and a number of these enzymes may be localized at both the light and electron microscope levels.

### Standard coupling azo dye method for acid phosphatase (Grogg and Pearse, 1952)

Fix tissue slices (2—4 mm thick) or slide cultures in 10 per cent neutral formalin at $4°$ C for 18 hr.

Where applicable cut frozen sections (10–15 $\mu$m) and mount on glass slides, to which they stick without adhesive, and dry for 1–2 hr at room temperature.

Incubate tissues at 37° C in: 20 ml of 0.1 M-acetate buffer pH 5.0 containing 20 mg sodium $\alpha$-naphthyl phosphate (Sigma Chemical Co.; Koch-Light Laboratories Ltd.), and 1.5 g polyvinyl pyrrolidone (stir well). Add 20 mg of stable diazotate of *o*-amino azotoluene (Fast Garnet GBC salt, Sigma Chemical Co.; Koch-Light Laboratories Ltd.), stir well and filter the mixture through GP paper onto the dry sections.

Incubate animal tissues 30–60 min.; plant and fungus material 1–2 hr; protozoa are variable, ½–3 hr.

Sites of acid phosphatase activity are brown with this particular diazonium salt.

*Note.* Many variations of the azo dye technique have been published involving many coupling agents. It is also possible to post-couple the reaction product to the diazonium salt in a separate solution in cases where the dyes inhibit enzyme activity. For variations see Pearse (1968).

(*b*) **Methods for other lysosomal hydrolases.** These are legion and usually depend upon variations of the azo dye technique. Several naphthol-containing substrates are now readily available (Koch-Light Laboratories Ltd.) which permit the localization of, e.g. peptidases, $\beta$-D-galactosidase, $\beta$-D-glucosidase, $\beta$-D-glucuronidase, $\alpha$-D-mannosidase, $\beta$-D-xylosidase, esterase, sulphatases, *N*-acetyl-$\beta$-D-glucosaminidase, $\beta$-L-fucosidase.

(*c*) **Ultrastructural localization of lysosomal hydrolases.** Methods for studying the ultrastructure of lysosomes are well established (*see* Daems *et al.*, 1972). The Gomori lead salt technique for acid phosphatase was the first histochemical method to be adapted for use with the electron microscope (Holt and Hicks, 1961). A modified version of this has now become a standard procedure in lysosome work.

*Gomori procedure for the localization of acid phosphatase at the level of ultrastructure*

Fix tissue blocks (1 mm cubes) in 3 per cent (w/v) glutaraldehyde in 0.1 M-cacodylate buffer pH 7.4 for 18 hr at 0°–4° C.

Wash blocks in 0.88 M-sucrose containing 1 per cent (w/v) gum acacia for 18 hr at 0°–4° C.

Wash for a few seconds in water.

Cut sections (40 $\mu$m) on a tissue chopper (or a cryostat).

Wash in 0.1 M-acetate buffer pH 5.0 (2 min).

Stain 20–30 min. using standard Gomori procedure (omitting sulphide treatment).

Wash several times (1 min. each) in water.

Post-fix in 1 per cent (w/v) osmium tetroxide solution (1 hr at 0°–4° C).

Dehydrate in an ethanol-propylene oxide series.

Embed in 'Araldite'.

Section on ultramicrotome.

Examine without post-staining in first instance.

Electron dense deposits of lead phosphate indicate sites of enzyme activity.

Rigid controls are necessary.

Post staining may be done, in order to obtain contrast, using lead citrate or uranyl acetate. It may be that Epon or Spurr resins are more suitable for plant and fungus materials.

These materials may also require longer periods of initial fixation in glutaraldehyde (up to 18 hr), and post-fixation of up to 4 hr in osmium tetroxide solution.

### Azo dye techniques for ultrastructural localization of hydrolases

Azo dye methods can in theory be used to locate hydrolases at the E.M. level, but in practice two main difficulties are encountered (*a*) the reaction products may be soluble in embedding fluids, (*b*) dye products show low electron opacity. Experimentation has proceeded but progress has been slow. However, some azo dye products are sufficiently insoluble to be useful and in addition electron density has been improved by synthesizing new diazonium compounds which contain, e.g. heavy metals. The technique of Smith and Fishman (1969) who used diazotized acetoxymercuric aniline (Taab Laboratories) in a post-coupling procedure for β-glucuronidase and acid phosphatase has yielded excellent results with liver, kidney and preputial glands of rats. Furthermore, reaction of the red pigment with thiocarbohydrazide (Taab Laboratories) appeared to form a stable mercury–sulphur linkage which rendered the final dye product insoluble in ethanol and embedding media. We have obtained good results with this method on animal, plant and fungus materials in our laboratory using the light microscope, but so far results have been disappointing with the electron microscope. Hislop *et al.* (1974) have obtained good ultrastructural localization of some lysosomal hydrolases in fungi using this method (see Chapter 2, Figs. 2.8, 2.9).

A high resolution technique for the fine-structural localization of acid hydrolases in tissues of rat and slug has been devised by Bowen (1971) using substituted naphthol substrates and both simultaneous and post-coupling with *p*-nitrobenzene diazonium tetrafluoroborate (Eastman Kodak Co.). However, in our hands the technique has proved of limited value with plant materials since even using a wide range of substituted naphthol substrates the various azo dye chromogens were soluble in both 'Araldite' and 'Durcupan' embedding media used for electron microscopy.

## 2. Techniques for observing lysosomes in living cells

(*a*) **Phase contrast microscopy.** Phase-dense refractile granules are readily observed in living tissues, especially in coverslip cultures of animal cells. Some of these particles may be equated with lysosomes, but since a variety of sub-cellular structures show similar properties this technique is of restricted value in lysosome work.

(*b*) **Vital staining.** Observations made last century by Pfeffler on plant cells and by Ehrlich on animal cells revealed that living cells take up and accumulate dyes in a variety of granules. Considerable evidence is now available that neutral red is taken into cells in non-toxic quantities and is concentrated within the lysosome system of cells of plants, animals and protists. This technique provides a useful method for observing lysosomes in living cells.

*Neutral red method for lysosomes*:  Incubate cells or tissues in a 1 in 20 000 solution of neutral red in 0.85 per cent (w/v) saline for 2 hr in darkness at 37° C for animal tissues and 20°–25° C for plants and protists. It is also possible to show acid phosphatase is located in some of these granules by post-fixing the stained cells for 10 min. in 6 per cent (w/v) glutaraldehyde in 0.1 M-cacodylate buffer (pH 7.4), washing in saline and staining for acid phosphatase by the standard coupling azo dye method. That a proportion of neutral red granules are also positive for acid phosphatase may be seen if the progress of staining for the latter is observed continuously with the light microscope.

(*c*) **Vital staining with fluorescent dyes.** Renewed interest in vital staining of lysosomes followed the observations of Koenig (1963) and Robbins, Marcus and Gonatus (1964) that the lysosomes of *living* cells accumulate acridine orange and related dyes and fluoresce under appropriate conditions in the fluorescence microscope. A full account of lysosome staining by fluorochromes is given by Allison and Young (1969).

*Fluorescence method for lysosomes*:  Useful for coverslip cultures of cells; protists in suspension or grown on coverslips; suspensions of plant tissue culture cells; suspensions of isolated organelles. Euchrysine 3R (Edward Gurr Ltd., London) is made up as a 1 in 1 000 (w/v) stock solution in normal saline (or culture medium, as appropriate).

Stain animal tissues for 15 min. to 2 hr at 37° C in a *final* dye concentration of 1 in $10^4$ (w/v) in darkness. Plant materials and some protists with cell walls require a stronger final dye concentration (1 in $5 \times 10^4$)

with incubation at 25° C in darkness. Wash in saline (centrifuge and resuspend suspensions). Mount in fresh saline (or medium, as appropriate). Examine in a fluorescence microscope with either transmitted or incident light.

A quartz iodine lamp is necessary (e.g. the 'Conference' microscope of Messrs. Gillett and Sibert) or a high pressure mercury vapour lamp (e.g. E. Leitz lamphouse 250 with HBO 200W lamp) with strong emissions in the blue-violet region (400—500 nm). Such lamps are used in conjunction with BG 12 excitation filters (2—4 mm thick) and an appropriate suppression filter (yellow minus blue; e.g. Ilford series Nos. 104—110 or Leitz K490) situated between the objective and eyepiece, preferably near the intermediate lens beneath the binocular head. With mercury vapour lamps it is necessary to incorporate, in addition, a red suppression filter, e.g. BG 38 and a heat absorbing filter near the lamp. Achromatic or fluorite objectives are satisfactory. Observations need to be made rapidly since the orange fluorescence of the lysosomes fades rapidly following photosensitization damage.

Many modifications and elaborations are reported in respect of illumination and optics for fluorescence microscopy, but the simple technique outlined above rarely gives results which are less than pyrotechnic. An improvement worthy of mention however, is the use of the incident illuminator based on the design of Ploem (E. Leitz) which improves the intensity of fluorescence against a dark background. Photograph using Kodak High Speed Ektachrome film (daylight type).

## Isolation of Lysosomes

1. From animal tissue

Difficulties in isolating lysosomes are related to their fragility, relative rarity and their diverse physical properties. The classic method is that devised by de Duve *et al.* (1955) for rat liver lysosomes:

Kill rats by a blow on the head and then bleed.

Remove the liver quickly and immerse in ice-cold medium (usually 0.25 M-sucrose) and allow to chill for a few minutes.

Take 1 vol. of thin slices of liver and homogenize in 3 vol. medium using a Potter-Elvehjem type homogenizer (a smooth glass tube with a close fitting Teflon pestle with *c*. 0.5 mm clearance) revolving at 1 300 rev/min. in ice at 0° C. A single upwards stroke is used to transfer the contents past the rotating pestle.

Optimal conditions are determined by trial and error.

Centrifuge slurry at 1 000 g for 10 min. (0° C) and treat the sediment in the homogenizer a second time. Spin at 1 000 g for 10 min. and combine the supernatant fluid fractions from the two extractions. Pellet by centrifugation at 18 000 g for 20 min. at 0° C. This pellet may be used as a crude lysosome preparation or it may be subjected to further purification, after resuspension in 0.25 M-sucrose solution, by one of the published schemes for differential centrifugation (de Duve, 1955; Slater, 1974) or by isopycnic centrifugation on either discontinuous or linear sucrose gradients. Generally linear gradients of buffered sucrose solutions are used to separate the lysosomes in cell homogenates from the other organelles contained therein. On occasions improved resolution may be obtained by using sugar alcohols, e.g. sorbitol and mannitol, or polysaccharides, e.g. Ficoll and glycogen in place of sucrose.

A large degree of variability is inherent in techniques for isolating subcellular organelles which makes the procedure an art rather than a precise science. Also a bewildering array of modifications of isolation techniques have been published for different tissues depending upon the nature of the problems involved and the particular equipment available in the laboratories in question. For those wishing to enter this field of endeavour the article of Beaufay (1972) is essential reading.

## 2. Lysosomes from plants

Techniques for isolating lysosomes from plants are even more variable than those used for animal tissues. One of the earliest successes is attributable to Matile *et al.* (1965) who isolated lysosomes from tobacco and maize:

Seeds of maize and tobacco are grown for three days at 27° C (use surface sterilized seeds).

Grind whole seedlings in a mortar and pestle in the presence of sand using 1 vol. of ice-cold 20 per cent (w/v) sucrose in 0.1 M-tris-HCl buffer (pH 7.1) containing lm M-EDTA.

Filter brei through cheesecloth.

Centrifuge 10 min. at 500 g to remove cell debris.

Layer supernatant fluid fraction onto a linear sucrose gradient (20–50 per cent sucrose containing lm M-EDTA).

Centrifuge 2–4 hr at 25 000 g.

Lysosomes equilibrate at a density of 1.22 g cm$^{-3}$.

Owing to the presence of potent oxidizing systems in many homogenates of plant materials it is often necessary to use a reducing buffer during homogenization. Pitt and Galpin (1973) and Pitt (1973) successfully isolated lysosomes from potato sprouts and potato leaves, respectively, using the following procedure:

*Extraction medium*, 0.1 M-tris-HCl buffer, pH 7.1, containing lm M-EDTA, 0.1 per cent (w/v) cysteine hydrochloride, 0.1 per cent (v/v) β-mercaptoethanol, 2.5 per cent (w/v) Ficoll (Pharmacia Fine Chemicals, Uppsala), 8.5 per cent (w/v) sucrose.

Grind tissues (1 g fresh wt./1 g extraction medium) for 2–3 min. in a mortar and pestle in the presence of sand at 0° C.

Strain through cheesecloth and centrifuge the brei at 1 000 g for 10 min. at 0°–4° C. The turbid supernatant may be used as a crude lysosome suspension or these organelles may be spun down for 30 min. at 18 000 g at 0°–4° C.

Further purification may be achieved by layering the turbid supernatant fluid onto a linear gradient (23 ml) made from extraction media containing 3.5 and 28.5 per cent (w/v) Ficoll. Spin in a swing-out head for 3 hr at 65 000 g. Lysosomal fractions equilibrate at 1.10 g cm$^{-3}$ and 1.07 g cm$^{-3}$ using potato sprouts. The procedure has been successful with potato leaves, pea roots and some other plant materials. In these instances the equilibrium densities are different from the values for potato sprout lysosomes.

### 3. Lysosomes from protists

In spite of the obvious attributes of many members of the Protista which make them attractive for lysosome studies few reports have appeared which record the successful isolation of this organelle from the group. The fungi have been examined in our laboratory, and with others we have encountered difficulties in the sub-cellular fractionation of these organisms. Only recently have the slime moulds (Ashworth and Wiener, 1973), the zooflagellates (Eeckhout, 1973) and the phytoflagellates (Aaronson, 1973) been submitted to extensive observations, and isolation techniques for the lysosomes of these groups are now being developed.

Most success has been achieved in the sub-cellular fractionation of protozoa with pioneering work being done by Müller's group. Generally the protozoon in question is grown in axenic culture and harvested by centrifugation. The pellet is resuspended in 0.32 M-sucrose buffered with 0.1 M-tris-HCl at, pH 7.2, containing 10mM-EDTA, and homogenized in a glass homogenizer of the Potter-Elvehjem type. Homogenates are fractionated by ultracentrifugation using swing-out or zonal rotors on either discontinuous or linear gradients of sucrose solution with a density range of 1.10–1.30 g cm$^{-3}$ (c. 15–60 per cent, w/v, sucrose). Successful fractionation procedures are exemplified by the work of Müller et al. (1966) and Lloyd et al. (1971) on *Tetrahymena*.

## Characterization of Isolated Lysosomes

Progress in the study of lysosomes was greatly aided by the introduction of the histochemical definition which complemented the initial rather rigid biochemical definition relating to this organelle. It is now fully realized that both morphological and biochemical criteria have limitations, consequently these are applied jointly whenever possible.

(a) **Morphological criteria.** The diversity of morphology of the components of the lysosome system has been alluded to many times in this book. It is this heterogeneity of form which limits the value of morphological criteria in judging the purity of lysosome preparations. The light microscope is of limited value in morphological studies of isolated lysosomes and the ultrastructural appearance in the electron microscope emphasizes this diversity of form. It is now generally recognized that a combination of cytochemical staining for acid hydrolases and ultrastructural observations is necessary to gain a precise knowledge of the purity of lysosome preparations. It is possible to fix suspensions of purified lysosomes (e.g. from density gradients) in a final concentration of 3 per cent (w/v) glutaraldehyde in 0.1 M-cacodylate buffer, usually containing isotonic sucrose, with subsequent embedding and thin sectioning of pieces of the pellet obtained on centrifugation. Direct fixation of pieces of a pellet is also possible. The usual cytochemical methods for lysosomal enzymes may be coupled with morphological observations. Care is needed in ultrastructural characterization of pelletted materials since stratification occurs during centrifugation and extensive observations of many pieces of a number of pellets may be necessary to convey an accurate overall impression of the organelle content.

(*b*) **Biochemical criteria.** The techniques for the biochemical assays of acid hydrolases are well defined and allow precise evaluation of the lysosomal hydrolase content of sub-cellular fractions within certain limits. It should be remembered that enzymes may be released from organelles which are damaged during the isolation procedure and these can become adsorbed onto cellular debris. Furthermore enzyme activity may be inhibited by components derived from the system under examination. Generally however, biochemical analysis of isolated fractions is the most reliable guide to the organelle content. When combined with morphological examinations of the ultrastructure and cytochemistry of isolated components it provides a powerful tool in sub-cellular fractionation studies. Unfortunately few tissues are amenable to this combined analytical approach.

Assays for lysosomal enzymes may be found in many appropriate texts on enzymology or via the extensive reviews of Tappel (1969) and Barrett (1972).

Mention of trade names in no way implies preference or recommendation over other commercial products of a similar nature.

---

# References quoted

AARONSON, S. (1973) Digestion in phytoflagellates, in *Frontiers of Biology*, 29. *Lysosomes in Biology and Pathology*, vol. 3, ch. 2, pp. 18–37, ed. J. T. Dingle. North-Holland Publishing Co., Amsterdam and London.
ALLISON, A. C. and YOUNG, M. R. (1969) Vital staining and fluorescence microscopy of lysosomes, in *Frontiers of Biology*, 14B. *Lysosomes in Biology and*

*Pathology*, vol. 2, ch. 22, pp. 600—28, eds. J. T. Dingle and H. B. Fell. North-Holland Publishing Co., Amsterdam and London.

ASHWORTH, J. M. and WIENER, E. (1973) The lysosomes of the cellular slime mould *Dictyostelium discoideum*, in *Frontiers of Biology*, 29. *Lysosomes in Biology and Pathology*, vol. 3, ch. 3, pp. 38—48, ed. J. T. Dingle. North-Holland Publishing Co., Amsterdam and London.

BARRETT, A. J. (1972) Lysosomal enzymes, in *Lysosomes — a Laboratory Handbook*, ch. 2, pp. 46—135, ed. J. T. Dingle. North-Holland Publishing Co., Amsterdam and London.

BEAUFAY, H. (1972) Methods for the isolation of lysosomes, in *Lysosomes — a Laboratory Handbook*, ch. 1. pp. 1—45, ed. J. T. Dingle. North-Holland Publishing Co., Amsterdam and London.

BOWEN, I. D. (1971) A high resolution technique for the fine-structural localization of acid hydrolases, *J. Microsc.*, 94, 25—38.

CARTLEDGE, T. G., COOPER, R. A. and LLOYD, D. (1971) Sub-cellular fractionation of eukaryotic microorganisms, in *Separations with Zonal Rotors*, ed. E. Reid. University of Surrey, Guildford.

DAEMS, W. Th., WISSE, E. and BREDEROO, P. (1972) Electron microscopy of the vacuolar apparatus, in *Lysosomes — a Laboratory Handbook*, ch. 4, pp. 150—99, ed. J. T. Dingle. North-Holland Publishing Co., Amsterdam and London.

de DUVE, C., PRESSMAN, B. C., GIANETTO, R., WATTIAUX, R. and APPELMANS, F. (1955) Tissue fractionation studies, 6. Intracellular distribution patterns of enzymes in rat liver tissue, *Biochem. J.*, 60, 604—17.

EECKHOUT, Y. (1973) Digestion and lysosomes in zooflagellates, in *Frontiers of Biology*, 29. *Lysosomes in Biology and Pathology*, vol. 3, ch. 1, pp. 3—17, ed. J. T. Dingle. North-Holland Publishing Co., Amsterdam and London.

GROGG, E. and PEARSE, A. G. E. (1952) A critical study of the techniques for acid phosphatase with a description of an azo dye method, *J. Path. Bact.*, 64, 627—36.

HISLOP, E. C., BARNABY, V. M., SHELLIS, C. and LABORDA, F. (1974) Localization of α-L-arabinofuranosidase and acid phosphatase in mycelium of *Sclerotinia fructigena*, *J. gen. Microbiol.*, 81, 79—99.

HOLT, S. R. and HICKS, R. M. (1961) The localization of acid phosphatase in rat liver cells as revealed by combined cytochemical staining and electron microscopy, *J. biophys. biochem. Cytol.*, 11, 47—66.

KOENIG, H. (1963) Intravital staining of lysosomes by basic dyes and metallic ions, *J. Cell. Biol.*, 19, 87A.

LLOYD, D. BRIGHTWELL, R., VENABLES, S. E., ROACH, G. I. and TURNER, G. (1971) Sub-cellular fractionation of *Tetrahymena pyriformis* ST. by zonal centrifugation: changes in activities and distribution of enzymes during the growth cycle and on starvation, *J. gen. Microbiol.*, 65, 209—23.

MATILE, P., BALZ, J. P., SEMADENI, E. and JOST, M. (1965) Isolation of spherosomes with lysosome characteristics from seedlings, *Z. Naturw.*, 20b, 693—8.

MÜLLER, M., BAUDHUIN, P. and de DUVE, C. (1966) Lysosomes in *Tetrahymena pyriformis*, *J. Cell Physiol.*, 68, 165—75.

PEARSE, A. G. E. (1968) *Histochemistry, Theoretical and Applied*, vol. 1, 3rd edition. J. & A. Churchill, London.

PITT, D. (1973) Solubilization of molecular forms of lysosomal acid phosphatase of *Solanum tuberosum* L. leaves during infection by *Phytophthora infestans* (Mont.) de Bary, *J. gen. Microbiol.*, 72, 117—25.

PITT, D. and GALPIN, M. (1973) Isolation and properties of lysosomes from dark-grown potato shoots, *Planta (Berl.)*, 109, 233—58.

ROBBINS, E., MARCUS, P. I. and GONATUS, N. K. (1964) Dynamics of acridine orange-cell interaction. II. Dye-induced ultrastructural changes in multi-vesicular bodies (acridine orange particles), *J. Cell. Biol.*, 21, 49—62.

SLATER, T. F. (1974) Lysosomes (with a short note on peroxisomes), in *Companion to Biochemistry*, ch. 17, pp. 511–51, eds. A. T. Bull, J. R. Lagnado, J. O. Thomas and K. F. Tipton. Longman, London.

SMITH, R. E. and FISHMAN, W. H. (1969) p-Acetoxymercuric aniline diazotate, a reagent for visualizing the naphthol AS-B1 product of acid hydrolase action at the level of the light and electron microscope, *J. Histochem. Cytochem.*, 17, 1–22.

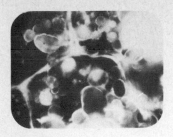

# Index

Acetoxymercuric aniline, 33, 149
β-N-acetylglucosaminidase, 10, 28, 122
  histochemistry, 148
Acid deoxyribonuclease, see deoxyribonuclease
Acid hydrolases:
  activation, 4, 24, 138
  barley seed, 114—19
  cytochemistry, 7, 8, 9, 146—9
  E.M. localization, 149
  extralysosomal, 5, 6, 139—42
  in development, 95
  inhibitors of synthesis, 104, 116
  injury and, 3, 110, 126—7, 138
  lysosomal, 4
  mammary gland involution, 98—100
  metamorphosis, 100—7
  pH optima, 2
  plant development, 108—19
  release in disease, 135—42
  solubilization, 18—20, 104, 139
  sponge development, 107—8
  uterus involution, 96—8
Acid phosphatase:
  classic lysosome marker, 3
  ER, 10
  extralysosomal locations, 5, 10
  Golgi apparatus, 10
  histochemistry, 146—9
  molecular forms, 5, 41—2
Acid ribonuclease, see ribonuclease

Acridine dyes:
  fluorescence of lysosomes, 25
  uptake by fungi, 29, 33
  uptake by Tetrahymena, 28
Actinomycin D:
  causing autophagy, 72
  preventing hydrolase synthesis,
    104—5
Activation, of lysosomal enzymes, 3
Adenosine triphosphatase (ATP-ase), 19
Adrenaline, 80
Adrenocortical trophic hormone
  (ACTH), 82
Aging, and lysosomes in Tetrahymena,
  71
Agriolimax reticulatus, lysosomes, 37
Agrobacterium tumefaciens, 142
Aleurone:
  barley, 117
  cotton, 115
  endosperm digestion, 114—19
  grains, 115
  in cotyledons, 117
  layer, 117
  lysosomes of, 117
Algae, lysosomes, 31, 34
Ambilysosomes, 66, 68, 73, 74, 77
  of mouse kidney, 68
Aminopeptidase, 120—1
Amoeba, lysosomes, 26
Amphibians, metamorphosis, 100, 104

α-amylase, 41, 114, 118
  germinating barley, 114
  plant lysosomes and, 41
  synthesis in barley, 116
Animal groups investigated for lyso-
    somes, 36
*Antheraea pernyi*, lysosomes and meta-
    morphosis, 103
Antibiotics:
  causing autophagy, 72
  inhibiting protein synthesis, 104—5
Antibodies, relations with lysosomes, 53,
    62
*Arachis hypogea*, spherosomes in, 47—8
Arthritis, lysosomes and, 133
Arylsulphatase, 4, 38, 146
Asbestosis, 129—30
*Aspergillus oryzae*, 91
*Asplenium fontanum*, sedimentable
    hydrolases, 42
Autoimmune response, in disease, 134
Autolysis:
  *Coprinus*, 122
  corolla, 111—13
  plant disease, 136
Autophagic body, *see* autophagic
    vacuole
Autophagic vacuole, 55, 66
  aging and, 71
  animal, 14—15, 64—9, 72—4
  formation, 72—5
  glucagon induction of, 73
  lysosome system and, 74
  origin of membranes, 72—4
  plants, 42, 67, 75
  protozoa, 75
  relations with ER, 75
  relations with Golgi body, 75
  sponge development, 108
  turnover and, 69
Autophagy:
  adverse conditions and, 69
  anuran metamorphosis, 106—7
  disease, 131
  enzyme changes in, 70
  factors causing, 72
  higher plants, 71
  induction of, 72, 74
  involution, 72, 96
  mammary gland, 99
  mechanisms of, 72
  metamorphosis, 72, 100
  pathology, 72
  protozoa, 69, 70, 71
  relations with heterophagy, 77

remodelling, 72
  starvation and, 72
  storage diseases, 128—31
  using $^{125}$I-labelled albumin to study,
    58
Autoradiography, of pinocytic vesicles,
    58
Azo dye method, 10
  acid phosphatase, 39, 147—8
  fungi, 31
  hydrolases, 148
  plants, 39, 42
  ultrastructural, 33, 149

Bacteria:
  acid hydrolases and, 35
  toxins, 17, 127
Balanced parasitism, 143
3,4-benzpyrene, stain for lysosomes, 25
Blood cells:
  histiocytes, 98
  lysosomes, 36, 98
  monocytes, 53, 98
  neutrophils, 53
Bone, 4, 86
  extracellular digestion of, 89
  intracellular digestion of, 89
  lysosomes of, 86, 89
  resorption of, 86, 88
  two-stage degradation of, 89
Bone marrow, lysosomes, 36
*Botrytis cinerea*, lysosomes, 30—3

Cartilage, 89
  erosion in disease, 133—4
  hydrolysis by cathepsin D, 85—6
Catalase, in *Tetrahymena*, 29
Catecholamine hormones and lysosomes,
    80
Cathepsin, 3, 4, 105
  in disease, 131
Cathepsin D:
  antigen, 86
  arthritis, 134
  connective tissue catabolism, 86
  immunochemistry, 86, 134
  immunological inactivation of, 86, 134
  lysosomal, 86
Cellular defaecation, 76, 77, *see also*
    exocytosis
*Chaos chaos*, 28
*Chara*, multivesicular bodies, 34
Characterization of lysosomes:
  morphological, 153
  biochemical, 154

*Index*

Chitinase, *Coprinus* autolysis and, 123
*Chlamydomonas*, phosphate regulation, 34
Chloramphenicol, effect on *Tetrahymena* lysosomes, 28
*Chlorogonium*, dictyosomes and vacuoles, 34
Chromaffin:
  cells, 80
  hormones, 80
Collagen, 4
  digestion of, 88
  disease and, 133–4
  fibrogenesis, 130
Collagenase, 4, 86
  lysosomal, 87
Connective tissue, 85
  cathepsin D and hydrolysis of, 86
  catabolism, 85
  resorption of, 86
  lysosomes and catabolism, 86
*Coprinus lagopus*, lysosomes and spore discharge, 123
Corolla wilting, role of lysosomes, 111–13
Cotton seed, 117
  aleurone grains, 115
Crinophagy, 77, 83
Crown gall, lysosomes in, 142
Cycloheximide, 104
  in autophagy, 72
Cytochemical methods for acid phosphatase, 7, 146–50
  liver, 8
  kidney, 9, 11
Cytolysome, 55
Cytosegresome, 55

*Dactylorhiza purpurella*, 143–4
  lysosomes in, 144
Damage:
  due to dusts, 129
  hydrolase changes in, 110
  to lysosomes, 2, 3
  to tissue, 128
de Duve, C., 2
Definitions of lysosomes, 24
Dense bodies, 15, 16
  in plants, 75
  in protozoa, 57
Density gradient ultracentrifugation:
  for lysosomes, 8, 151–2
  plant lysosomes, 42–7
  *Tetrahymena*, 28, 29

Deoxyribonuclease, 3, 4, 38, 105, 112–13, 128
Detergents:
  ambilysosome formation, 66
  effect on lysosomes, 2, 12, 17, 38, 66, 97
  heterophagy, 66
Development of flowers, lysosomes and, 111–14
Diaminobenzidine (DAB) method, 60
  peroxidase, 58–9
Diauxic growth:
  lag phase, 120
  lysosomes and, 119–21
Diazonium salts, 148, *see* azo dye
*Dictyostelium discoideum*, autophagy in, 28
Differential centrifugation, lysosomes, 152
*Dionoea*, digestive enzymes in, 93
Disease:
  animal lysosomes and, 128–35
  definition, 126
*Drosophila*, 36

Ecdysone, 100
  gene activation by, 107
Ecdysterone, induction of secondary lysosomes, 102
Electron dense markers:
  carbon, 25
  colloidal metals, 25
  thorium dioxide, 25
Endocrine system, 83
  regulation and modulation, 80
Endocytosis, 5, 12, 52, 77
  bone resorption, 88
  disease therapy, 133
  peroxidase marking, 12
Endoplasmic reticulum (ER), 1
  autophagic bodies, 75
  origin of lysosomes, 77
Endosperm, digestion by hydrolases, 114–19
*Ephydatia mulleri*, 107
ER, *see* endoplasmic reticulum
*Escherichia coli*, 36, 64, 119
Esterase, 4, 41, 114, 135
  histochemistry, 148
Euchrysine, fluorescence stain for lysosomes, 25, 150
*Euglena gracilis*, 16, 52
  cytochemistry, 35
  origin of lysosomes, 34

159

Exocytosis, 76, 77
  defaecation, 76
  extracellular digestion, 76
  regurgitation, 77
  reversed pinocytosis, 76, 79
Extracellular defaecation, 76–7
Extracellular digestion, 52, 76, 85–9
  acid hydrolases and, 90
  by fungi, 89
  lysosomes in, 85
Extracellular secretion:
  acid hydrolases, 89
  by fungi, 28, 89
  in insectivorous plants, 93
Extralysosomal hydrolases, 5, 10
  in endoplasmic reticulum, 10
  in Golgi apparatus, 10
Extraorganism hydrolase secretion:
  by insectivorous plants, 93
  in fungi, 89–93
  in higher plants, 93–5

Flowers, lysosomes in development,
  110–14
Fluorescent antibody technique, for
  lysosomal enzymes, 25, 86, 134
Fluorescence microscopy:
  acridine orange, 25
  3,4-benzpyrene, 25
  euchrysine 3R, 33, 150
  lysosomes, 25, 29, 150–1
Food vacuoles:
  flagellates, 52
  protozoa, 52
  slime moulds, 52
Freezing, effect on lysosomes, 2
Fruits, lysosomes in, 114
α-fucosidase, in disease, 131
  cytochemistry, 148
Fungus:
  acid hydrolases, 91–2
  α-L-arabinofuranosidase, 91
  β-galactosidase, 91, 93
  lysosomes, 119–23
  necrotrophy, 89
  origin of lysosomes, 91–2
  parasitism, 89
  saprophytism, 89
  secretion of hydrolases, 91
*Fusarium caeruleum*, 136
  dry rot of potatoes, 136–7

β-galactosidase, 4, 17, 93, 134, 146
  histochemistry, 148
  in fungi, 29

in plants, 38
Gargoylism, 131
Genetic diseases, lysosomes and, 130–3
GERL hypothesis of lysosome origin,
  13–15
Germination of barley, and acid hydrolases, 114–19
Gibberellic acid (GA₃), 114
  barley germination, 114
  protein synthesis, 116
Glucagon, induction of autophagy by,
  72–3
1,3-β-glucanase, of barley, 114
Glucose-6-phosphatase, ER membrane
  marker, 73, 74
α-glucosidase, 4, 17, 28, 131
β-glucosidase, 122
  histochemistry, 148
  in *Tetrahymena*, 28
β-glucuronidase, 3, 4, 19, 38, 146
  histochemistry, 148
β-glycerophosphatase, 9, 10, 147
  in root development, 109
  plant, 41
Glycogen, 4
  overloading of cells, 131
Glycogenosis, 131
Glyoxysome, marker enzymes, 41–2
Golgi apparatus, 1
  origin of lysosomes, 12–16, 56,
  74–7
Gomori, G., 1
Gomori technique, 7, 9, 11, 24–5
  acid phosphatase in fungi, 30
  E.M. procedure, 148
  in plants, 40–1
  showing structure-linked latency, 25
Gout, 130
Growth, definition, 95
Growth regulating hormone, 82
Heterophagy: *see also* endocytosis
  autoradiography of, 60–1
  bacteria, 62
  plants, 53, 64
  relations with lysosome system, 77
  relations with autophagy, 77
Higher plants:
  acid hydrolases secreted by, 93
  lysosomes in, 93
Histochemistry, of lysosomes, 146–54
Hormones:
  catecholamine, 80
  effect on lysosomes, 97
  reproductive, 97
  thyroid, 80

Hormone secretion:
  crinophagy, 77
  lysosome functions in, 79
  modulation of, 79, 83
  pituitary, 80
  regulation, 80—1
Hurler's syndrome, 131
Hyaluronidase, 4
Hydrocortisone, 17, 63, 89
  disease therapy and, 135
  stabilizers of lysosome membrane, 63
Hydrolase:
  azo dye methods, 148
  barley endosperm digestion, 114—19
  *de novo* synthesis, 117
  latency, 2
Hydrolytic enzymes: *see* hydrolases and
  acid hydrolases
Hypersensitivity, in plants, 54, 141—2

Immunochemical assay of hydrolases,
  137—8
Inborn genetic disorders:
  lysosomal enzymes, 131—3
  storage diseases, 76, 129—31
Indoxyl esterases, in fungi, 29
Inflammation, and lysosomes, 130—5
Injury, and lysosomes, 127—8
Insectivorous plants:
  absorption of protein by, 94
  acid hydrolases and digestion in, 93
Intracellular digestion, 4, 12
  lysosomes and, 52—77
Involution:
  mammary, 98—100
  uterus, 96—8
*Ipomoea purpurea* (Morning glory),
  hydrolases and development,
  111—14
Isolation of lysosomes, 7, 8, 151—2
  animals, 151
  plants, 152
  protists, 153

Juvenile hormone, 100

Kidney lysosomes, cytochemical
  location, 9, 58, 60, 68

Labelled amino acids:
  incorporation in tadpole, 105
  metamorphosis, 105
Labilizers, of lysosomes, 17, 97
Lactogenic hormone, 82, 83, *see also*
  mammotrophic hormone

modulation of by lysosomes, 83
Latency, of lysosomal enzymes, 17, 18,
  24
Lead salt method, *see* Gomori technique
  acid phosphatase, 146
Leaves, lysosomes, 110
Lecithinase, *see* phospholipase, 2
*Leishmania*, 28
L-fraction, 2
Lipochondria, 1
Lipofuscin, in residual bodies, 75, 129
Liposomes, 133
Lipoxygenase, 38
Liver lysosomes, cytochemical localization, 9
Luteinizing hormone, 82
Lysosomal enzymes, table of, 4
Lysosome:
  autophagy and, 64—72
  biochemical definition, 24
  concept, 2—5
  cytochemistry of, 7, 146—9
  damage to, 3, 126—45
  discovery, 2
  disease, 3, 126—45
  disruption, 3
  early model, 3
  extracellular digestion, 52, 85
  fate of, 14
  heterogeneity, 4, 54
  heterophagy, 55—64, 77
  hormone secretion, 80
  in development, 95—123
  interrelationships of components, 15
  intracellular digestion, 52—78
  latency of hydrolases, 17, 24
  membrane, 17—22
  methods of study, 7, 146—54
  origin, 12—16, 74, 77
  polymorphism, 54
  system, 1—6
Lysozyme, 4
Lytic bodies, 2

Macrophage:
  in disease, 129—30
  involution and, 97—9
  lysosomes of, 36
  metamorphosis, 106—7
Mammary gland:
  involution, 98
  lysosomes, 98—100
Mammotrophic hormone (*see* lactogenic
  hormone), 83
  modulation of by lysosomes, 83

Mammotrophs, 82
α-mannosidase, 4, 17
  histochemistry, 148
Membrane of lysosomes:
  enzyme locations in, 17—18
  disruption of, 2, 3, 17, 19
  hypotheses on structure, 19—22
  labilizers, 17, 97
  origin of, 12—16, 17—22, 74, 77
  stabilizers, 17
  steroid hormones, 97—8
Menstrual cycle, lysosomes and, 96—7
Metachromasia of lysosomes, use of
  toluidine blue, 25
Metamorphosis:
  anuran, 105—6
  hydrolase synthesis *de novo* in, 100
  insect, 100—4
Methods, for lysosome study, 146—54
Microscopy, of lysosomes, 146—51
Microsomes, 1
Microtubules, intracellular transport and,
  77
Mitochondria, 1, 2, 8, 28, 127
  effect of lysosomes on, 127
  in autophagic vacuoles, 64, 66, 68
  markers for, 44
  rate of turnover, 69
Mouse spleen macrophages, lysosomes
  in, 21
Mucopolysaccharides, 4
  in disease, 131
  in lysosomes, 25
Multivesicular body, 14, 34
  origin, 55
Mycoplasma, 64
Mycorrhiza, 142—3
  lysosomes, 143
Myelin figures, 82, 99, 106

Nerve tissues, lysosomes in, 36
*Neurospora crassa*, 28, 36, 91
Neutral red staining, 127
  lysosomes, 29, 150
  uptake by *Paramecium*, 26
Noradrenaline, 80
Novikoff, A. B., 2
Nucleases, 4, 28
  *Tetrahymena*, 28, 29

Oestradiol, lysosome damage, 97—8
Oestrogen, effect on lysosome activity,
  97
Onion bulb-scale, lysosomes in, 38
*Ophryoglena*, lysosomes, 26

Orchid, lysosomes, 144
Origin of lysosomes, 14—16
Osmoregulation:
  in *Poecilia*, 85
  lysosomes and, 84—5
Osmotic shock, and lysosome disruption,
  2
Osteoblast, 87
Osteoclast, 87, 89
Osteocyte, 87

*Paramecium caudatum*:
  acid phosphatase, by ultrastructural
    method, 27
  azo dye method, 27
  digestion, 56—8
  lysosomes, 26, 57
Peptidase, 4
  histochemistry, 148
Periodic-acid-Schiff (PAS), for lysosome
  polysaccharides, 25
Peroxidase:
  accumulation of by kidney lysosomes,
    5
  phagosome marker, 12, 58
  technique, 25, 58—9
Peroxisomes, 28, 30
  half-life, 69
Phagocyte, 53, 54, 62
Phagocytosis, 12, 52
  diaminobenzidene method, 58
  discovery of, 53
  electron dense markers, 60
  gout, 130
  peroxidase technique, 58
  in plants, 53
  in protozoa, 26, 52
Phago-lysosome (*see* phagosome), 52, 58
Phagosome, 52, 54, 77
  detection by autoradiography, 58—60
  gold-DNA marker, 60
  peroxidase method, 58, 60
  phago-lysosome, 58
Phase contrast microscopy, of lysosomes, 150
Phosphatase, alkaline, Gomori procedure
  for, 1
Phosphodiesterase, acid, 4, 41, 104
Phospholipase, 4, 135
  lysosome disruption, 2, 38
  release of latent hydrolases, 17—18
Phospholipase A, release of lysosomal
  enzymes by, 20
Phospholipids, 4
  in lysosomes, 25

Photosensitization, damage to lysosomes, 128, 151
Phytoalexins, in plants, 54
*Phytophthora erythroseptica*, potato pathogen, 135–8
*Phytophthora infestans*, 110
  disease, 139–41
  potato blight, 136
  release of particulate hydrolases, 139–40
*Pinguicula*, digestive enzymes of, 93
Pink-rot disease of potatoes, 135–8
Pinocytic vesicles:
  $^{125}$ I-labelling, 58–9
  micropinocytic vesicles, 25
  peroxidase technique, 58–9
  relations with lysosomes, 58–9
Pinocytosis, 12, 26, 52, 77
  demonstration in kidney using $^{125}$ I, 58–9
  in plants, 44, 53, 64, 111
  relation with lysosome system, 58–9
*Pisum sativum*, 39, 40, 67, 109, 117, 153
Pituitary gland:
  endocrine system, 80
  hypothalamus, 80
  pituitary hormones, 80
  tadpole, 100
Plant disease, and lysosomes, 135–42
*Poecilia latipinna*, lysosomes, 84
Pollen grain, lysosomes, 39, 110
*Polytomella caeca*, particulate hydrolases, 35
Pompe's disease, *see* glycogenosis, 131–2
Potato blight, 136, 139–42
  lysosome disruption in, 139
Potato shoot lysosomes, 18, 19, 20
  isolation of, 6, 43–4
  molecular forms of phosphatase in, 6
Potato tuber diseases, lysosomes in, 135–40
Pregnancy, lysosomes in, 99
Primary lysosomes, 13, 74, 81, 88
  ER and Golgi, 13–15
  in protozoa, 13–16
  interrelations with lysosome system, 55
  origin, 55
Prolactin (lactogenic hormone):
  lysosomes and, 82
  in *Poecilia*, 84
Prostate gland, lysosomes in, 36
Protease, 17, 41, 114, 120–2, 135

in fungi, 28
in plants, 41, 93
in *Tetrahymena*, 28–9
Protein synthesis, 111
  autophagy and, 72
  *de novo*, in tadpole, 104–5
  in regression, 104
  metamorphosis and, 104
Prothoracic gland, in insects, 100
Protozoa, lysosomes in, 26, 28
Pseudo-gout, 130
Puromycin, 104
  causing autophagy, 72
*Pythium ultimum*:
  Golgi origin of secretory vesicles, 30
  membrane differentiation, 92
  secretory bodies in, 91–3

Radiations, causing autophagy, 17, 72
Regression, 85–8
  bone, 88
  connective tissue, 85
  mammary gland, 98–9
  metamorphosis, 100–7
  uterus, 96–8
Residual body, 99
  aging and, 75
  Golgi apparatus, 14, 64
  origin, 55
  relation with lysosome system, 15
  smooth ER and, 14, 64
Reticulo-endothelial system (RE):
  components of, 54
  lysosomes in cells of, 36
Retinol (vitamin A alcohol), 20
Reversed pinocytosis (*see* exocytosis), 76
  cell regurgitation, 77
  defaecation, 76–7
  exocytosis, 79
*Rheovirus*:
  and lysosomes, 63
  mode of infection, 63
*Rhizoctonia solani*, 143–4
Ribonuclease, 3, 4, 18, 91, 110, 114, 120, 121
  activation in plant disease, 135–8
  plant, 41, 93, 135–8
  virus infection and, 63
Ribonucleic acid (RNA):
  metamorphosis, 105
  synthesis, 4, 63
Root development, and lysosomes, 108–9

*Saccharomyces cerevisiae*:
  diauxic growth and hydrolases,
    119—20
  lysosomes, 30
*Sarcophaga bullata*, ecdysterone and
  lysosomes, 101—2
*Sclerotinia fructigena*, 33, 91
Secondary lysosomes, 14, 74, 81, 88
  interrelations with lysosome system, 55
  metamorphosis, 101—2
  origin, 55
Secretion of hydrolases:
  by fungi, 90—1
  extracellular, 85—9
  extraorganism, 90, 93
  in barley germination, 116
  in higher plants, 93—4
Secretory granules:
  hormones and, 77
  lysosomes and, 77
Seeds, lysosomes of, 114—19
Silicosis, 129—30
Silkworm, oak, metamorphosis in,
    103—4
Slime-moulds, lysosome-like organelles
  of, 28
Somatotrophic hormone, lysosomes and,
    82
Spermatozoa, lysosomes in, 36
Spherosomes, 45—8
*Spirodela oligorrhiza*, phosphatase in
  phosphate deficient, 71
Spleen, lysosomes in, 36
Sponges, lysosomes and vitelline plate-
  lets, 108
Stabilizers, of lysosomes, 17, 63
Stems:
  fractionation of potato, 41—3
  lysosomes in, 19, 20, 44, 46, 109—10
*Stentor*, lysosomes, 26
Storage diseases:
  asbestosis, 129—30
  gout, 130
  lysosomes and, 129—33
  silicosis, 129—30
Streptolysins, effect on lysosomes, 62,
    127
Substrates, of lysosomal enzymes, 2, 4

Techniques for lysosomes, 146—54
*Tetrahymena pyriformis*, 16
  aging, 71
  autophagy, 69
  changes in lysosomes in starvation,
    70—1

distribution of marker enzymes, 29
  enzymic characteristics of lysosome
    populations, 28
  hydrolases and growth, 119
  starvation, 72
  uptake of acridine dyes, 28
*Thanatephorus cucumeris*, 143—4
Thioacetic acid esterase, in fungi, 32
Thyroglobulin, hydrolysis by lysosomes,
    80—1
Thyroid gland, 81
  lysosomes in, 36
Thyroid hormones, lysosomes and secre-
    tion of, 80
Thyroid stimulating hormone (TSH),
    80—1
  in anuran metamorphosis, 100
Thyrotrophic hormone, amphibian, 100
Thyrotrophs, 82, 83
Thyroxine, 79, 80, 100
Tissue:
  damage by lysosome contents, 128
  regression, 85—8
Tomato fruit, lysosomes, 114
Toxins, bacterial, 17, 127
*Trichomonas*, 28
Tri-iodothyronine, 79, 100
  tail regression in tadpoles, 105
Triton WR-1339:
  accumulation in lysosomes, 12
  use in isolating 'pure' lysosomes, 12
Triton X-100, effects on lysosome
  membranes, 38
*Trypanosoma*, 28
TSH, *see* thyroid stimulating hormone,
    80—2
Tumours, and lysosomes, 130
Turnover, 53, 69
  and half-life of organelles, 69
  of cell components, 69

Uterus:
  involution, 96
  lysosomes in, 96—8

Vacuoles:
  as lysosome equivalents, 43—4, 111
  food, 52
  fungus, 29
  plant, 16, 40, 43, 93, 110
*Verticillium albo-atrum*, 91
*Vicia faba*, 109
  lysosomes, 39
Virus:
  lysosomes during infection by, 63

picornia, 63
  *Rheovirus*, 63
Virus uncoating, lysosomes and, 63
Vital staining, of lysosomes, 25, 150
Vitamin A, 17, 85
  in cartilage breakdown, 85
  labilizer of lysosomes, 89

*Xenopus*, tail regression in, 106
$\beta$-xylosidase, histochemistry, 148

*Zea mays*, acid phosphatase localization
  in, 41, 109, 152
Zooflagellates, 28, 153
Zymogen, synthesis by pancreas, 13, 82